Quantum Mechanics and the Philosophy of Alfred North Whitehead

AMERICAN PHILOSOPHY SERIES
Douglas R. Anderson and Jude Jones, series editors

QUANTUM MECHANICS

and the Philosophy
of Alfred North Whitehead

by

MICHAEL EPPERSON

Fordham University Press
New York • 2004

Cover art image: "The Lorenz Attractor in 3D" by Paul Bourke. Swinburne University of Technology (used with permission)

American Philosophy Series, No. 14
ISSN 1073-2764

Library of Congress Cataloging-in-Publication Data

Epperson, Michael.
    Quantum mechanics and the philosophy of Alfred North Whitehead / by Michael Epperson.—1st ed.
        p.   cm. — (American philosophy series, ISSN 1073-2764 ; no. 14)
    Includes bibliographical references and index.
    ISBN 0-8232-2319-1
    1. Quantum theory—Philosophy.   2. Whitehead, Alfred North, 1861–1947. I. Title. II. Series.
QC174.13.E66   2004
530.12'01—dc22   2003023756

Printed in the United States of America
08 07 06 05 04 5 4 3 2 1
First edition

Στην μνημη της αγαπημενης μου
Για γιας
Ελενη Ξυρουχακης Δοκιμακης
(1910–1999)

CONTENTS

PREFACE

WHATEVER ONE MAY SAY about the unsurpassed predictive power of quantum mechanics, few would argue that it is a more comfortable or intuitive theory than the classical mechanics of Newton and Galileo, which its innovators intended to replace as *die endgültige Physik*. For while we have for the past several hundred years enjoyed classical mechanics both in application of its predictive power and in contemplation of its descriptive power, quantum mechanics, though certainly providing vast improvements in the accurate prediction of phenomena, does so only in deficit of its ability to describe these phenomena intuitively.

A coherent and intuitive characterization of nature, such as that given us in classical mechanics and its underlying ontology of mechanistic materialism, has been sorely lacking in quantum mechanics. One reason is that many of its earliest innovators, Einstein, Planck, and Bohr among them, had presumed that quantum mechanics could be accommodated by the same classical ontology of fundamental materialism, with perhaps a few minor modifications, such that efforts toward a novel ontology were for many years thought unnecessary. But such an accommodation has, after several decades of work, proven to be an infamously uneasy one as evinced by the many notorious quantum-classical incompatibilities and "paradoxes" that have unfortunately become the defining characteristic of quantum mechanics for many.

One need look no further than the familiar problem of wave-particle duality to glimpse the difficulty. Quantum mechanics seems to entail two competing and incompatible fundamental descriptions of nature, and this leaves one with three alternatives: (i) to characterize nature as fundamentally particulate wherein wave-like properties are an abstraction; (ii) to characterize nature as fundamentally wave-like wherein particulate properties are an abstraction; (iii) to pass through these two horns and deny that nature is capable of fundamental characterization at all (apart from this sanction itself, of course) such that we merely characterize our complementary *experiences* of nature as wave-like or particle-like depending on the circumstances, rather than characterizing nature herself.

To each of these three viewpoints we can associate various theorists—Einstein, for example, to the first, Schrödinger to the second, Bohr to the third, and so forth—and we can trace the many various subsequent mediations of these three viewpoints back to a commitment to one against the others. The statistical interpretation of Born is such an example, wherein the wave-like aspects of nature operative in quantum mechanics are interpreted as statistical probability amplitudes pertaining to the largely unknowable positions, momenta, or other qualifications of particles. Such a mediation preserves the epistemic sanctioning of the third alternative—the admission of an epistemic veil that shrouds nature just enough that she cannot be known with complete deterministic certainty in every qualification; it preserves the viewpoint of the second alternative, such that the wave-like characterization of nature, interpreted as a probability amplitude, describes the precise transparency of this veil; and it preserves the viewpoint of the first alternative such that what lies beneath the veil—nature herself—is characterized as fundamentally particulate and deterministic. For Born, "events happen indeed in a strictly causal way, but . . . we do not know the initial state exactly. In this sense the law of causality is therefore empty; physics is in the nature of the case indeterminate, and therefore the affair of statistics."[1]

Of course, wave-particle duality as it pertains to quantum mechanics entails a host of other related difficulties, each of which has been similarly attended to by various theorists according to the three aforementioned viewpoints. For example: nature described as fundamentally (and classically) fluid and deterministic according to the first viewpoint; nature described as fundamentally discontinuous and probabilistic according to the second viewpoint; our experiences of nature described as either fluidly deterministic or discontinuously probabilistic, depending on the circumstances, according to the third viewpoint. The failure to accommodate quantum mechanics adequately according to a classical materialist ontology—that is, solely according to the first viewpoint described earlier—is especially evident in the phenomenon of nonlocal causal interrelations predicted by quantum mechanics.

---

[1] Max Born, *Atomic Physics, 8th ed.* (New York: Dover, 1989), 102.

As with the interpretation of Born, which entailed a mediation from among the three viewpoints, one finds a similar mediation in the nonlocal hidden variables interpretation of David Bohm. Certain quantum mechanical predictions, confirmed by experiment, entail nonclassical interactions between two spatially well separated systems such that a measurement performed upon one system instantly (and therefore nonclassically) affects measurement outcomes in the other system. A deterministic interpretation of quantum nonlocality would therefore seem to require a violation of special relativity, as if some sort of superluminal influence were transmitted from one system to the other. Bohm's mediation among the three alternative viewpoints mirrors that of Born when applied to an interpretation of this phenomenon, with the addition of a causally efficacious "pilot-wave" thought to propagate superluminally through an ether-like medium of point-particles. By the operation of this pilot-wave, the two spatially well separated systems are, despite appearances, fundamentally non-separate beneath the veil of epistemic uncertainty caused by this ether of point-particles—particles whose qualities are incapable of complete deterministic qualification and thus "hidden."

We have, in this and other interpretations, attempts to equip quantum mechanics with a descriptive power comparable in strength to its predictive power; but in the attempt to produce a characterization of nature as predictably and descriptively satisfying as that given by classical mechanics, these and other interpretations so contort the very classical fundamental materialism that they attempt to preserve that one tends to feel even less satisfied with these "classical" interpretations of quantum mechanics than one felt with no ontological interpretation at all. One cannot be surprised, then, by the widespread appeal of sheer instrumentalism when it comes to interpreting quantum mechanics in the classroom and laboratory, where the best interpretation is often held to be no interpretation at all.

There is, however, a fourth viewpoint which, for many physicists and philosophers, provides the key to a coherent and intuitive interpretation of quantum mechanics. This viewpoint begins with the understanding that formally, quantum mechanics describes nothing more than the evolution of a system of facts from an initial state to a final state, where the term "state" refers to a maximal specification of the facts belonging to the system measured. Further, the outcome state yielded by quantum mechanical prediction is not a singular state, but rather a matrix of probable states among which one will become actualized in accord with its probability valuation. This unique actualization is not accounted for by quantum mechanics, but it is anticipated by the mechanics and is confirmed retrodictively upon subsequent observation. In other words, actual initial facts give rise to sets of potential facts that evolve to become actual final fact in a quantum mechanical measurement interaction. Here, the terms "evolution" and "probability" both presuppose an actuality prior to the evolution and anticipate an actuality subsequent to the evolution: The expressions "X evolves to become Y" and "0.5 is the probability that X will become Y" reflect this presupposition and anticipation. Since these actualities are presupposed and anticipated by quantum mechanics, in the same way that matter is presupposed and anticipated by classical mechanics, quantum mechanics cannot be used to account for the existence of actualities any more than classical mechanics can be used to account for the existence of matter. The essence of quantum mechanics, then, lies not in the qualification of what exists before and after measurement as emphasized by the classical materialistic ontology—an ontology of being, where reality is identified with actuality; the essence of quantum mechanics, rather, is the evolution itself—an ontology of becoming, where reality is seen to comprise two fundamental species: actuality and potentiality—"first principles" in that each is incapable of abstraction from the other.

Heisenberg suggested a re-adoption of this Aristotelian concept of potentia as a means toward a coherent interpretation of quantum mechanics, and later theorists including Robert Griffiths, Murray Gell-Mann, James Hartle, and Roland Omnès, among others, have incorporated this notion of potentia into the concept of "histories" of quantum mechanical evolutions. A macroscopic material object thus becomes characterized most fundamentally as a history of evolutions of discrete facts or events—evolutions from actuality to potentiality to actuality. The problem of coherently interpreting quantum nonlocality, among other problems, is thus easily solved: As a particularly dramatic example, one can intuitively understand how the history describing the ongoing evolution of an atom or molecule or person or nation is instantaneously and "nonlocally" affected by an asteroid that has just been knocked by a comet into a collision course with Earth, with impact to occur in two years. Although Earth and the asteroid are spatially well separated, the newly evolved actuality pertaining to the asteroid's course change has instantaneously affected the potentia associated with the ongoing histories describing Earth and any fact associated with it. Since quantum mechanical histories are in a perpetual state of augmentation, quantum event by quantum event, at any such event the potentia associated with a history condition its future augmentation in a way similar to (but not identical to) the way antecedent facts condition a history's future augmentation.

In other words, as the potentia associated with a history change with even a single quantum event, the history itself changes, as does the system defined by the history. Although we on Earth cannot be causally *influenced* by the asteroid's course change sooner than the time it takes for a photon to travel from the asteroid to Earth, the potentia associated with any physical system on Earth have clearly been causally *affected*, and affected instantly and nonlocally (though not determined). In the same way, it has been demonstrated experimentally that the phenomenon of quantum nonlocality cannot be used for faster-than-light communication. This limitation is understandable and entirely intuitive according to this interpretation, since such communication would entail nonlocal causal influence of actualization rather than merely nonlocal causal affection of potentia as described in the example above.

This distinction between causal affection of potentia and causal influence of actualization is just one of many conceptual innovations inherent in the interpretation of quantum mechanics according to an event ontology—an ontology of historically evolving process—rather than an ontology of fundamental mechanistic materialism. Many theorists have gone on to show other advantages of such an interpretation, including how it is able to account for the one-way direction of time in thermodynamics as ontologically rather than merely epistemically significant—a concept whose compatibility with other interpretations of quantum mechanics is problematic at best. But the primary advantage is that it is an interpretation that defies any "quantum-classical" dualism, such that classical mechanics becomes an abstraction from the more fundamental quantum mechanical description of nature, rather than merely a complementary and incompatible description. And unlike the proposals of Born and Bohm, among others, this interpretation requires neither an ar-

bitrary epistemic sanction in the form of "hidden variables" nor selective violations of the very classical mechanics these proposals were
intended to preserve.

There are, however, a great many ontological innovations and implications inherent in the accommodation of quantum mechanics by
a metaphysics grounded in this idea of historically evolving process,
and these require a careful systematic exploration that is likely to
exceed the purview and interests of most physicists. It is therefore
both fortunate and remarkable that one finds in the philosophy of
Alfred North Whitehead—developed in its most systematic form
during the same years that brought the quantum theoretical innovations of Bohr, Born, Schrödinger, Heisenberg, Dirac, et al—a
metaphysical scheme that so precisely mirrors the hypothetical deductions and inductions made by the physicists who have contributed to the development of the event-ontological, "historical
process" interpretations of quantum mechanics. The purpose of this
essay, then, is to point out and explicate these correlations and their
significance to the interpretation of quantum mechanics, and more
broadly, to the philosophy of science in general; for Whitehead's
repudiation of fundamental materialism and Cartesian dualism
echoes loudly in the work of recent theorists such as Robert Griffiths,
Roland Omnès, Wojciech Żurek, and Murray Gell-Mann, among
several others, whose own repudiation of fundamental materialism
and "quantum-classical" dualism is the most recent attempt to solve
a philosophical problem whose roots extend all the way back to the
problem of χωρισμός introduced by Plato—the supposed chasm
separating what is from what appears to be.

I am much indebted to the following people for their invaluable
and generous advice, comments, criticisms, and instruction: John R.
Albright of the Department of Physics at Purdue University; John B.
Cobb, Jr. and David Ray Griffin of the Claremont Graduate Univer-
sity; Peter E. Hodgson, head of the Nuclear Physics Theoretical
Group of the Nuclear Physics Laboratory of the University of Oxford;
Timothy E. Eastman, Group Manager for Space Science, NASA
Goddard Space Flight Center; Hank Keeton; and especially David
Tracy and Franklin Gamwell of the University of Chicago.

M.G.E.

University of Chicago
July 2003

1

Introduction

This chapter is intended to provide a brief overview of the synthesis developed over the course of the book. As a result, it occasionally incorporates certain concepts and terminology that have yet to be introduced. Since this book was written for readers with varying familiarity with quantum mechanics and Whitehead's philosophy—including no familiarity with either—readers with some knowledge of both should begin with this chapter, whereas those who need familiarization with the subjects might skip ahead to chapter 2.

THE ATTEMPTED CORRELATION of quantum mechanics and Whitehead's cosmological scheme—or any philosophical scheme, for that matter—is an endeavor to be expected of both philosophers and physicists discomfited by the various "paradoxical" conceptual innovations inherent in quantum mechanics when interpreted according to the classical ontology of mechanistic materialism. That various proposed correlations of quantum mechanics and Whitehead's cosmology have come from both philosophers and physicists, then, should not surprise, nor should their respective emphases of approach: The philosophers tend to depict the physical side of the correlation in overly broad strokes in order to avoid the infamously complicated concepts and terminology inherent in quantum mechanics, and the physicists, who prefer to avoid the infamously complicated concepts and terminology inherent in Whiteheadian cosmology, tend to depict his metaphysical scheme in similarly broad strokes.

Some of the proposals made thus far—those suggested by Abner Shimony,[1] Henry Folse,[2] and George Lucas,[3] for example—have proven useful in establishing an initial dialogue; but they have tended to break down once a certain level of detail is approached, on either the physics side or the philosophy side. With respect to the latter, the reason lies not in any failure by philosophers to comprehend quantum mechanics adequately, but rather with the advocacy of certain popular interpretations of quantum mechanics founded upon and inspired by concepts wholly incompatible with the Whiteheadian cosmological scheme. These incompatibilities are most easily evinced by the extent to which a particular interpretation of quantum mechanics fails to meet the four desiderata Whitehead requires of his and any philosophical interpretation of experience—physical, microphysical, or otherwise. Such an interpretation, writes Whitehead, should be: (i) *coherent*, in the sense that its fundamental concepts are mutually implicative and thus incapable of abstraction from each other; (ii) *logical*, in the ordinary sense of the word, as regards consistency, lack of contradiction, and the like; (iii) *applicable*, meaning that the interpretation must apply to certain types of experience; (iv) *adequate*, in the sense that there are no types of experience conceivable that would be incapable of accommodation by the interpretation.[4]

Thus, for example, attempts to demonstrate the compatibility of Bohr's principle of complementarity and Whiteheadian metaphysics, though perhaps useful in terms of higher-order epistemological issues, fails for lack of coherence at the most fundamental level, the very level for which it was intended. Bohr's two complementary characterizations of our experiences of nature—classical and quantum—are not mutually implicative, and this is the very point of complementarity. Henry Folse suggests that a correlation of Bohr's interpretation of quantum mechanics and Whitehead's philosophy is in order, primarily because of the repudiation of fundamental mechanistic materialism common to both; "however," Folse admonishes, "the fate of any potential alliance is in jeopardy so long as current discussions of the subject insist on concentrating on the fine points of quantum interpretation rather than its broader more general ramifications." He continues:

> Quite naturally there are many aspects of the philosophy of organism which find no counterpart in the philosophical extrapolations of the Copenhagen Interpretation. . . . There is no reference to the equivalents of "feeling," "satisfaction," or "conceptual prehension." Yet Whitehead would have anticipated this, for the physicists' interpretation of theory is based on a very small segment of experience; Whitehead's system aims at far greater compass.[5]

The difficulty is that concepts like "feeling," "satisfaction," and "conceptual prehension" are fundamental to Whiteheadian metaphysics. They are not higher-order abstractions that should be, or even can be, ignored whenever applied to the specialized interpretation of physical experiences. But aside from specific correlatives in the physical sciences for the terms "feeling," "satisfaction," and "conceptual prehension," which Whitehead does, in fact, specify,[6] the incompatibility of Bohr's interpretation of quantum mechanics and Whitehead's metaphysical scheme lies most fundamentally in the simple failure of Bohr's principle of complementarity to meet the desideratum of ontological coherence.

Similar attempts to ally Whitehead's cosmology with David Bohm's nonlocal hidden variables interpretation of quantum mechanics fail for the same reason, despite the focus upon certain significant compatibilities, such as that of (i) Bohm's "implicate order" pertaining to the etherlike field of all actualities in the universe, correlate with (ii) the analogous concept of necessarily and mutually interrelated actualities in Whitehead's scheme, as well as the repudiation of fundamental classical "extended substance" common to both. In Bohm's scheme, however, the repudiation of fundamental substance (Bohm's particles, though concrete, are more akin to Einstein's "point-instants" and Whitehead's "actual occasions" than extended substance) is not a repudiation of deterministic, mechanistic materialism, as it is in Whitehead's ontology. Bohm's fundamentally deterministic "implicate order" inherent in the field of all actualities entails symmetrical and therefore purely deterministic relations among these actualities.

Insofar as these relations remain hidden within the deep realm of Bohm's "implicate order," our participation in this order is restricted to manifold epistemically limited observational contexts. Bohm suggests that because of this, his theory in no way vitiates conceptions of freedom, creativity, novelty, and so forth—principles central to Whiteheadian metaphysics. However, given that the fundamental implicate order of the universe is deterministic, hidden though this order may be, it is difficult to see how freedom grounded in epistemic ambiguity can be thought to be as significant as freedom grounded in an ontological principle—even if our finite observational contexts all but guarantee such ambiguity. Bohm writes:

> As long as we restrict ourselves to some finite structures of this kind,
> however extended and deep they may be, then there is no question of

complete determinism. Each context has a certain ambiguity, which may, in part, be removed by combination with and inclusion within other contexts. . . . If we were to remove all ambiguity and uncertainty, however, creativity would no longer be possible.[7]

An ontologically significant principle of freedom from determinacy requires an asymmetrical temporal modality and its associated logical order, where the past is settled and closed and the future is open—a temporality that is irreversible. This is a key feature of Whitehead's metaphysics. Though Bohm's implicate order is fundamentally temporally symmetrical and deterministic, he suggests that there is some similarity between Whitehead's process of concrescence and the quantum mechanical relationships among the actualities of his "implicate order" cosmology. "A key difference," he notes,

> is that these relationships are grounded in the deeper, "timeless" implicate order that is common to all these moments. . . . It is this implicate "timeless" ground that is the basis of the oneness of the entire creative act. In this ground, the projection operator P_n, the earlier ones such as P_{n-1}, and the later ones such as P_{n+1} all interpenetrate, while yet remaining distinct (as represented by their invariant algebraic structures).[8]

Epistemic uncertainty as to the specifications of most of these relations manifests itself as the familiar, temporally asymmetrical "explicit order" characterizing our experiences, such that temporal priority appears reflective of logical priority. This reflection is evinced, for example, by the one-way direction of time associated with the laws of thermodynamics. But if one could peer through the epistemic veil of this temporal asymmetry—if one could perceive the implicate order of hidden variables and its associated "pre-space"—then the fundamentally symmetrical relationship among past, present, and future would be revealed. Bohm writes:

> If it were possible for consciousness somehow to reach a very deep level, for example, that of pre-space or beyond, then all "nows" would not only be similar—they would all be one and essentially the same. One could say that in its inward depths now *is* eternity, while in its outward features each "now" is different from the others. (But *eternity* means the depths of the implicate order, not the whole of the successive moments of time.)[9]

But since temporal priority is merely epistemically significant by such an interpretation, it is unclear how it might have any significant

correlation with an ontologically significant logical priority. As mentioned earlier, such a gulf between the contingent and the necessary
has its roots in the problem of χωρισμός, or "separation" of necessary forms from contingent facts in Plato's metaphysics. It is a problem central to many interpretations of quantum mechanics, and also
to interpretations of the special and general theories of relativity—
the latter with respect to the relationship between the formal geometrical character of spacetime and the facts constitutive of
spacetime. In the general theory of relativity, Einstein bridges Plato's
χωρισμός by deriving the formal geometry of spacetime from the
events themselves; this approach to χωρισμός, then, has a certain
compatibility with the hidden variables interpretations of quantum
mechanics discussed earlier. (The close relationship between quantum mechanics and theories of spatiotemporal extension is addressed at length in chapter 5.)

In the Whiteheadian cosmology, the integration of (i) the asymmetrical, logical modal relations among facts and (ii) the symmetrical, relativistic modal relations among spatiotemporal forms of facts,
is a function of the fundamental dipolarity of actualities. But in
Whitehead's scheme, the asymmetrical, logical ordering among actualities as genetically related, serially ordered becomings, is, in one
sense, the fundamental order upon which their symmetrical, relativistic spatiotemporal ordering is predicated. The existence of facts is
thus, by the requirement of logic, necessarily prior to their spatiotemporal ordering in Whitehead's metaphysical scheme. But Bohm's
hidden variables interpretation entails the opposite—that it is the
symmetrical, deterministic relations among actualities which are
fundamental to the asymmetrical—and by his interpretation, ontologically insignificant—logical ordering of the actualities themselves.
Thus, the irreversibility of thermodynamic processes, for example, is
by Bohm's interpretation merely a statistical epistemic artifact of an
underlying purely deterministic, symmetrical, "implicate" order.

Bohm and his colleague B. J. Hiley illustrate this fundamental deterministic symmetry of the implicate order by describing the workings of a particular experimental apparatus:

> This device consists of two concentric glass cylinders; the outer cylin-
> der is fixed, while the inner one is made to rotate slowly about its axis.
> In between the cylinders there is a viscous fluid, such as glycerine, and

into this fluid is inserted a droplet of insoluble ink. Let us now consider what happens to a small element of fluid as its inner radius moves faster than its outer radius. This element is slowly drawn out into a finer and finer thread. If there is ink in this element it will move with the fluid and will be drawn out together with it. What actually happens is that eventually the thread becomes so fine that the ink becomes invisible. However, if the inner cylinder is turned in the reverse direction, the parts of this thread will retrace their steps. (Because the viscosity is so high, diffusion can be neglected.) Eventually the whole thread comes together to reform the ink droplet and the latter suddenly emerges into view. If we continue to turn the cylinder in the same direction, it will be drawn out and become invisible once again.

When the ink droplet is drawn out, one is able to see no visible order in the fluid. Yet evidently there must be some order there since an arbitrary distribution of ink particles would not come back to a droplet. One can say that in some sense the ink droplet has been enfolded into the glycerine, from which it unfolds when the movement of the cylinder is reversed.

Of course if one were to analyse the movements of the ink particles in full detail, one would always see them following trajectories and therefore one could say that fundamentally the movement is described in an explicate order. Nevertheless within the context under discussion in which our perception does not follow the particles, we may say this device gives us an illustrative example of the implicate order. And from this we may be able to obtain some insight into how this order could be defined and developed.[10]

Bohm and Hiley go on to suggest that this implicate order "contains explicate suborders as aspects which are particular cases of the general notion of implicate order. In this way we clarify our earlier statement that the implicate order is general and necessary, while explicate orders are particular and contingent cases of this."[11]

The predication of actualities upon the relativistic spatiotemporal *relations* among actualities—the predication of facts upon their implicate ordering—similarly manifests itself in popular quantum cosmogonic models such as those proposed by Stephen Hawking and James Hartle, wherein a vacuous spacetime is purported to evolve quantum mechanically from a void of pure potentiality—potentiality somehow abstracted from actuality. Such a void, often termed a "quantum vacuum" or "quantum foam," is a fundamentally incoherent construction, given that the concept of actuality is necessarily presupposed by the concept of potentiality, such that the latter cannot be abstracted from the former. This is both a logical requirement and a requirement of quantum mechanics, which describes the evolution of actual facts and their associated potentia— not the evolution of vacuous potentia into actuality.

These conceptual impediments to the fundamental logic and coherence of the preceding interpretations of quantum mechanics all stem from a common source—the attempt to use quantum mechanics to account for the existence of actualities, when quantum mechanics both presupposes and anticipates their existence. This presupposition and anticipation is clearly reflected in the mathematical concept of probability, which—as it pertains to the termination of a quantum mechanical measurement in a matrix of probable actualities rather than a determined, unique actuality—is a quantifiable propensity that a *presupposed* fact will evolve to become a quantifiably *anticipated* novel fact. (In quantum mechanics, and in Whiteheadian metaphysics, the anticipated unique novel fact is both subsequent to *and* consequent of the evolution.) Any interpretation of quantum mechanics that meets the desideratum of logic, then, cannot include a quantum mechanical account of the existence of actualities, which are both presupposed and anticipated by the mechanics.

The two interpretations of quantum mechanics briefly described earlier—those of Bohr and Bohm—were both born of inductive philosophical generalizations, which is to be expected of scientific theories to some degree. But these generalizations, each in its own way, fail to meet one or more of the Whiteheadian desiderata for a sound philosophical scheme by which we can coherently and logically interpret our experiences of the physical world. "The only logical conclusion to be drawn, when a contradiction issues from a train of reasoning," writes Whitehead, "is that at least one of the premises involved in the inference is false."[12] As regards these two interpretations of quantum mechanics, the culprit premise is the concept of fundamental mechanistic materialism. Bohr attempts to salvage this concept by draining it, and its complementary quantum theoretical conception of nature, of all ontological significance; the facts of objective nature are thus permanently veiled to the extent that we must replace the notion of "objective facts of nature" with public coordinations of our experiences of nature.[13] And Bohm attempts to salvage the primacy of mechanistic materialism by resorting to a similar veil, such that the *apparent* openness of the future by its asymmetrical relations with the facts of the past—as related to the *apparent* indeterminacy of quantum mechanics, for example—is merely a statistical artifact of an epistemic handicap that prevents us from observing and specifying the ether of "hidden variables." This ether, for Bohm, constitutes the implicit, underlying universe of fundamentally symmetrically related facts—that is, a fundamentally deterministic universe.

In contrast, however, recent years have brought the development of a family of interpretations of quantum mechanics formulated in part as a response to these difficulties. This family of interpretations uses only the orthodox "Copenhagen" quantum theoretical formalism, but abstracted from the philosophical sanctions placed by Bohr upon its proper interpretation. Instead, it begins with a decidedly nonclassical concept, suggested by Heisenberg (and resurrected from Aristotle), that actuality and potentiality constitute two fundamental species of reality. This new characterization of potentia as ontologically significant by itself does much to eliminate the infamous paradoxes of quantum mechanics, as Heisenberg points out,[14] and it is an acute example of the importance, commended by Whitehead, of imaginative generalization in the construction of a sound philosophical scheme; for Heisenberg's characterization of potentia as ontologically significant picks up where the inductive generalizations from classical mechanics failed in their attempted logical and coherent application to the quantum theory. And coupled with the explicit acknowledgment that quantum mechanics cannot be used to account for the existence of actualities, which it necessarily presupposes and anticipates, these two concepts—actuality and potentiality as fundamental species of reality—form the cornerstone of this new family of interpretations. These interpretations characterize quantum mechanics not as a means of describing the actualization of potentia (for the terminal actuality, like the antecedent actuality, is presupposed by the mechanics), but rather as a means of describing the *valuation* of potentia. And as regards the central role of mathematics in these quantum mechanical valuations, Whitehead clearly believes that the success of specific mathematical concepts upon which the quantum theory is founded—probability, tensors, and matrices, to name a few—derives from their origination in the "imaginative impulse," controlled by the requirements of logic and coherence: "It is a remarkable characteristic of the history of thought that branches of mathematics, developed under the pure imaginative impulse, thus controlled, finally receive their important application. Time may be wanted. Conic sections had to wait for 1800 years. In more recent years, the theory of probability, the theory of tensors, the theory of matrices are cases in point."[15]

And as the imaginative impulse was central to the formulation of the mathematical concepts and formalism of quantum mechanics, so would it be to the formulation of a coherent and logical interpretation of quantum mechanics. Thus, from the imaginative conception of ontologically significant potentia, the speculative generalization is further expanded to include three more concepts— each of which is presupposed by the quantum formalism: (i) that actualities evolve to become novel actualities, forming historical routes of actualities, and it is their associated potentia which mediate this evolution from actuality to actuality; (ii) that the evolution of any actuality somehow entails relations with *all* actualities by virtue of the closed system, required by the Schrödinger equation, comprising all actualities; (iii) that these necessary relations, when relative to a single evolution, require a process of negative selection whereby the coherent multiplicity of relations is reduced into a set of decoherent, probability-valuated, mutually exclusive and exhaustive, potential novel integrations, such that superpositions of mutually interfering potentia incapable of integration are eliminated. This process of negative selection guarantees that histories of actualities are mutually consistent (that is, in compliance with the logical principles of non-contradiction and the excluded middle), such that the novelty of the future does not vitiate the actuality of the past.

This process of negative selection describes what is referred to as the "decoherence effect," and the family of interpretations referred to earlier consists of those that agree upon the ontological significance of this process, as well as those related concepts discussed earlier. Many notable theorists, including Robert Griffiths, Wojciech Żurek, Murray Gell-Mann, and Roland Omnès, among several others, have demonstrated that the interrelations among all facts— those belonging to a measured system, a measuring apparatus, and the environment englobing them—play a crucial conceptual and mechanical role in the elimination of superpositions of nonsensical, interfering potentia (and potential histories) via negative selection and the resulting decoherence effect. Though the proposals of each of these thinkers differ somewhat, they all emphasize the importance of this negative selection process. According to Żurek's Environmental Superselection interpretation, for example, "Decoherence results from a negative selection process that dynamically eliminates non-classical [i.e., mutually interfering and thus incompatible] states."[16] Żurek, like many physicists who believe in the central importance of this process of negative selection, maintains that decoherence is a consequence of the universe's role as the only truly closed system, which, put another way, guarantees the ineluctable "openness" of every subsystem within it. "This consequence of openness is critical in the interpretation of quantum theory," Żurek continues, "but seems to have gone unnoticed for a long time."[17]

The quantum mechanical evolution of the state of a system is thus characterized as the valuation of potential novel facts, and potential novel histories of facts, as the evolution proceeds relative to the historical route of actualities constitutive of itself and its universe. The valuations of potentia terminal of this quantum mechanical evolution are describable as a matrix of probabilities, such that they are mutually exclusive and exhaustive—that is, additive in the usual sense. Also, as probabilities, the actualities of the past are presupposed, and a unique actuality terminal of the evolution is similarly presupposed and anticipated. It is understood, then, that this final phase of the evolution—the unitary reduction to a single actuality—lies beyond the descriptive scope of quantum mechanics.

The evolutionary valuation of potentia in quantum mechanics can be correlated phase by phase, and concept by concept, with Whitehead's metaphysical scheme, such that the former can be characterized as the fundamental physical exemplification of the latter. Both entail that a world of mutually interrelated facts (Whitehead's "actual occasions") is presupposed and anticipated in the evolution of each novel fact, and that the inclusion of these facts of relatedness (Whitehead's "prehensions" of facts as "data") in the act of measurement, by these necessary mutual interrelations, somehow entails the following: (i) all other facts and their associated potentia—either in their inclusion in the specification, or their necessary exclusion from the specification. This requirement is reflected in Whitehead's "Principle of Relativity" and his "Ontological Principle," and in

quantum mechanics by the Schrödinger equation's exclusive applicability to closed systems, with the universe being the only such
system. The exclusions relate to the process of negative selection
productive of the decoherence effect, to be discussed presently, and
Whitehead refers to these eliminations as "negative prehensions."
Their form and function with respect to environmental degrees of
freedom are identical to those related to the process of decoherence;
(ii) the evolution of the system of all facts into a novel fact—namely,
a maximal specification (the "state" specification) of the relevant
facts (those not excluded by decoherence or "negatively prehended"
in Whitehead's terminology). State specification—the maximal
specification of many facts via the necessary exclusion of some
facts—thus entails the evolution of a novel fact—namely, a unification of the facts specified; and (iii) the requirement that this evolution proceed relative to a given fact, typically belonging to a
particular subsystem of facts. In quantum mechanics, these are, respectively, the "indexical eventuality" and the "measuring apparatus"; Whitehead's equivalent conceptions are, respectively, the
"prehending subject" and its "nexus"—that is, the system of actualities to which the subject belongs. This requirement that state evolution (Whiteheadian "concrescence") always proceed relative to a
particular fact or system of facts is given in Whitehead's "Ontological Principle" and "Category of Subjective Unity"; their correlates in
quantum mechanics—the necessary relation of a state evolution to
some preferred basis characteristic of the measuring apparatus—has
often been misapprehended as a principle of sheer subjectivity, the
source of the familiar lamentations that quantum mechanics destroys the objective reality of the world.

Measurement or state specification thus entails, at its heart, the
anticipated actualization (concrescence) of one novel potential fact/
entity from a matrix of many valuated potential facts/entities that
themselves arise from antecedent facts (data); and it is understood
that the quantum mechanical description of this evolution terminates in this matrix of probability valuations, anticipative of a final
unitary reduction to a single actuality. Ultimately, then, concres-
cence/state evolution is a unitary evolution, from actualities to
unique actuality. But when analyzed into subphases, both concrescence and quantum mechanical state evolution are more fundamentally nonunitary evolutions, analogous to von Neumann's conception

of quantum mechanics as most fundamentally a nonunitary state evolution productive of an anticipated unitary reduction.[18] It is an evolution from (i) a multiplicity—the actual many—to (ii) a matrix of potential "formal" (in the sense of applying a "form" to the facts) integrations or unifications of the many (Whitehead's term is propositional "transmutations" of the many—a specialized kind of "subjective form"—and he also groups these into "matrices"[19]). Each of these potential integrations is described in quantum mechanics as a projection of a vector representing the actual, evolving multiplicity of facts onto a vector representing a potential "formally integrated" outcome state (eigenstate). The Whiteheadian analog of the actual multiplicity's "projection" onto a potential integration is "ingression"—where a potential formal integration arises from the ingression of a specific "potentiality of definiteness"[20] via a "conceptual prehension" of that specific potentiality (Whitehead also refers to these potential facts as "eternal objects" and explicitly equates the two terms[21]). But whereas in quantum mechanics, the state vector representing the actual multiplicity of facts is projected onto the potential integration (the eigenvector representing the eigenstate), in Whitehead's scheme it is the latter which ingresses into the prehensions of the actual multiplicity. This difference reflects Whitehead's concern with the origin of these potentia, for if they ingress into the evolution, then by his Ontological Principle they must be thought of as coming from somewhere. The eigenstate, or object of projection in quantum mechanics, is, in contrast, simply extant, and indeed, this is one of the infamous philosophical difficulties of quantum mechanics.

There are, furthermore, two important characteristics shared by both the quantum mechanical and Whiteheadian notions of potentia that should be noted here. First, there is a sense in which both are "pure" potentia, referent to no specific actualities. For Whitehead, "eternal objects are the pure potentials of the universe; and the actual entities differ from each other in their realization of potentials."[22] "An eternal object is always a potentiality for actual entities; but in itself, as conceptually felt, it is neutral as to the fact of its physical ingression in any particular actual entity of the temporal world."[23] In quantum mechanics, this pure potentiality is reflected in the fact that the state vector $|\Psi\rangle$ can be expressed as the sum of an infinite number of vectors belonging to an infinite number of subspaces in an infinite number of dimensions, representing an infinite number of potential states or "potentialities of definiteness" referent to no specific actualities and potentially referent to all. Many of these are incapable of integration, forming nonsensical, interfering superpositions, and are eliminated as negative prehensions in a subsequent phase of concrescence.

Second, quantum mechanical potentia are also "inherited" from the facts constituting the initial state of the system (as well as the historical route of all antecedent states subsumed by the initial state) such that preferred bases in quantum mechanics are typically reproduced in the evolution from state to state. Similarly, in Whitehead's scheme, antecedent facts, when prehended, are often "objectified" according to one of their own historical "potential forms of definiteness"—typically, the given potential forms that were antecedently actualized at some point in the historical route of occasions constituting the system measured.

> An actual entity arises from decisions *for* it and by its very existence provides decisions *for* other actual entities which supersede it.[24]

> Some conformation is necessary as a basis of vector transition, whereby the past is synthesized with the present. The one eternal object in its two-way function, as a determinant of the datum and as a determinant of the subjective form, is thus relational. . . . An eternal object when it has ingression through its function of objectifying the actual world, so as to present the datum for prehension, is functioning "datively."[25]

Whitehead's characterization of potentia as "relational," then, is clearly exemplified by the manner in which potentia mediate the actuality of a measured system and the actuality of the outcome of the measurement—that is, the mediation between the initial and final system states.

The quantum mechanical state evolution/concrescence thus continues into its next phase: (iii) a reintegration of these integrations into a matrix of "qualified propositional" transmutations,[26] involving a process of negative selection where "negative prehensions" of potentia incapable of further integration are eliminated. The potential unifications or propositional transmutations in this reduced matrix are each qualified by various valuations. Each potential transmutation relative to the indexical eventuality of the measuring apparatus (i.e., each potential outcome state relative to the apparatus and some prehending subject belonging to it) is thus a potential "form" into which the potential facts of the universe will ultimately evolve. Whitehead terms these "subjective forms." As applied to quantum mechanics, the term *subjective* refers to the fact that the "form" of each potential outcome state is reflected in the preferred basis relative to the indexical eventuality of the measuring apparatus (i.e., the prehending subject). Again, it is only the form that is thus subjective—for any number of different devices with different preferred bases could be used to measure a given system, and any number of different people with their own "mental preferred bases" could interpret (measure) the different readings of the different devices, and so on down the von Neumann chain of actualizations. The potential facts to which each subjective form pertains, however, are initially "given" by the objective facts constitutive of the world antecedent to the concrescence at hand. Thus, again, the "subjective form" of a preferred basis is in no way demonstrative of sheer subjectivity—that is, the evolution of novel facts *as determined* by a particular subject. It is, rather, demonstrative of the evolution of novel facts jointly determined by both the world of facts antecedent to the evolution *and* the character of the subject prehending this evolution by virtue of its inclusion in it. According to Whitehead's "Ontological Principle,"

> Every condition to which the process of becoming conforms in any particular instance, has its reason either in the character of some actual entity in the actual world of that concrescence, or in the character of the subject which is in process of concrescence. . . .[27]

The actual world is the "objective content" of each new creation.[28]

The evolution thus proceeds to and terminates with what Whitehead terms the "satisfaction," which in quantum mechanical terms is described as (iv) the anticipated actualization of the final outcome state—that is, one subjective form from the reduced matrix of many subjective forms—in "satisfaction" of the probability valuations of the potential outcome states in the reduced matrix. In quantum mechanics, as in Whitehead's model, this actualization is irrelevant and transparent apart from its function as a datum (fact) in a subsequent measurement, such that the "prehending subject" becomes "prehended superject." Again, this is simply because both White-head's process of concrescence and quantum mechanics presuppose the existence of facts, and thus cannot account for them. For White-head, "satisfaction" entails "the notion of the 'entity as concrete' abstracted from the 'process of concrescence'; it is the outcome sepa-rated from the process, thereby losing the actuality of the atomic entity, which is both process and outcome."[29] Thus, the probability valuations of quantum mechanics describe probabilities that a given potential outcome state *will be actual* upon observation—implying a subsequent evolution and an interminable evolution of such evolu-tions. Every fact or system of facts in quantum mechanics, then, subsumes and implies both an initial state and a final state; there can be no state specification S without reference, implicit or explicit, to $S_{initial}$ and S_{final}. This is reflected in Whitehead's scheme by refer-ring to the "subject" as the "subject-superject":

> The "satisfaction" is the "superject" rather than the "substance" or the "subject." It closes up the entity; and yet is the superject adding its character to the creativity whereby there is a becoming of entities superseding the one in question.[30]

> An actual entity is to be conceived as both a subject presiding over its own immediacy of becoming, and a superject which is the atomic creature exercising its function of objective immortality. . . .[31]

> It is a subject-superject, and neither half of this description can for a moment be lost sight of. . . .[32]

> [The superject is that which] adds a determinate condition to the settlement for the future beyond itself.[33]

Thus, the process of concrescence is never terminated by actual-ization/satisfaction; it is, rather, both begun and concluded with it. The many facts and their associated potentia become one novel state (a novel fact), and are thus increased historically by one, so that, as Whitehead puts it, "the oneness of the universe, and the oneness of each element in the universe, repeat themselves to the crack of doom in the creative advance from creature to creature."[34] "The atomic actualities individually express the genetic unity of the universe. The world expands through recurrent unifications of itself, each, by the addition of itself, automatically recreating the multiplicity anew."[35]

The specific phase-by-phase, concept-by-concept correlation of Whitehead's cosmological scheme and the decoherence-based interpretations of quantum mechanics—such that the latter are seen as a fundamental physical exemplification of the former—satisfies Whitehead's intention that his cosmological model be compatible with modern theoretical physics. Indeed, much of the development of the "Copenhagen" quantum formalism occurred contemporaneously with Whitehead's development of his cosmological scheme. Whitehead writes: "The general principles of physics are exactly what we should expect as a specific exemplification of the metaphysics required by the philosophy of organism."[36] But it also satisfies the intention of the quantum theory's originators that it provide the fundamental physical characterization of nature—"*die endgültige Physik*"—an intention that can be fulfilled only within the context of a coherent, logical, applicable, and adequate ontological interpretation.[37]

Ultimately, the test of the synthesis proposed herein, as is the case for any adventure in speculative philosophy, is to be found in renewed observation mediated by the metaphysical scheme both in areas of physics and in other areas as well—areas that, apart from the scheme, might have seemed entirely unrelated.

> The success of the imaginative experiment is always to be tested by the applicability of its results beyond the restricted locus from which it originated. . . . The partially successful philosophic generalization will, if derived from physics, find applications in fields of experience beyond physics. It will enlighten observation in those remote fields, so that general principles can be discerned as in process of illustration, which in the absence of the imaginative generalization are obscured by their persistent exemplification.[38]

The study of complex adaptive systems in nature, as one such application, has been the topic of a great deal of research and debate over the past several years, and has significant roots in attempts by several physicists to demonstrate that quantum mechanics describes such complexity at the most fundamental physical level. The "balanced complexity"[39] described by Whitehead as the "subjective aim" governing the evolution of novel actuality in his cosmological scheme has a direct analog in the concept of "effective complexity"—also a balance of regularity (Whitehead's genetic "reproduction" of potentia) and diversity ("reversions" of potentia from the genetic regularity). Efforts have been made in the sciences to disclose the fundamental function and exemplification of effective complexity by referring to quantum mechanics, and the decoherence-based interpretations are particularly well suited to this task. The reasons are especially clear in the context of the Whiteheadian cosmology; for the decoherence effect is predicated upon the very notions of contrast of (i) *diverse* multiplicities of facts with (ii) *regulated* potential integrations of these facts (the regulation being a product, in part, of negative selection) into alternative, probability-valuated, mutually exclusive forms of definiteness.

The application of the decoherence-based interpretations of quantum mechanics to the study of complexity in nature, where the former is seen as a fundamental exemplification of the latter, is an area of inquiry significant not only to the philosophy of science but also, potentially, to the philosophy of religion. The contextualization of quantum mechanics in terms of the Whiteheadian cosmological scheme is commended here, for Whitehead's repudiation of fundamental mechanistic materialism is also a repudiation of its correlate characterization of the universe as a cold realm of mechanical accidents from which our purportedly illusory and sheerly subjective perceptions of purpose and meaning are, by certain views, thought to derive. In the words of Jacques Monod, the Nobel-laureate biochemist: "Man knows at last that he is alone in the universe's unfeeling immensity, out of which he emerged only by chance."[40] In sharp contrast, by Whitehead's cosmology as exemplified by the decoherence interpretations of quantum mechanics, the universe is instead characterized as a fundamentally complex domain with an inherent aim toward an ideal balance of reproduction and reversion—a balance formative of a nurturing home for a seemingly infinitely large family of complex adaptive systems such as ourselves.

The usefulness of the synthesis of quantum mechanics and Whitehead's cosmology to conversations among philosophy, science, and religion is further demonstrated as it might apply to the role of God as primordial actuality in quantum mechanical cosmogonic models of *creatio ex nihilo*, such as the one proposed by Stephen Hawking and James Hartle[41] mentioned earlier. Quantum mechanics describes the evolution of the state of a system of actualities always in terms of an initial state antecedent to the evolution, and a matrix of probable outcome states subsequent to and consequent of the evolution. Therefore, the application of quantum mechanics to the description of any cosmogonic model—an inflationary universe model, for example—still requires a set of "initial conditions" or initial actualities at $t = 0$, when the evolution begins. In such an evolution, there must logically be, in other words, some actuality which evolves. Renaming these initial conditions "quantum vacuum" or "quantum foam" (equating them mathematically to the empty set), despite the intended connotations of these linguistic and mathematical terms, does nothing to relieve the theory of its logical obligation to presuppose the existence of facts antecedent to and subsequent to (and consequent of) a quantum mechanical evolution—whether this evolution describes the emission of an X ray from a black hole, or the emission of the universe from a black foam. For without these initial actualities, there can be no spacetime structure in which a quantum mechanical state evolution might operate. Hawking's suggestion that it is a vacuous spacetime that first evolves quantum mechanically into actuality from sheer potentiality (i.e., from no initial actuality) defies the logically necessary predication of spacetime upon existence—the logically necessary predication of the ordering among actualities upon the actualities themselves.

But the Whiteheadian philosophy is likely to be useful in such conversations for another reason that has less to do with the facts of his philosophy than its form. For the spirit of speculative philosophy which animated both the development of Whitehead's cosmological scheme and that of the decoherence-based interpretations of quantum mechanics will be equally useful to the rapidly widening conversation among philosophy, science, and religion. An appeal to the Whiteheadian spirit of speculative philosophy would do much to mediate and advance such conversations; for theories would, in this spirit, have the character of philosophic generalization hypothetically deduced relative to careful scientific observations, but coupled with the play of a free imagination and conditioned by the requirements of coherence and logic. The product of this creative amalgam, applied to subsequent observations, propels the process forward with the explicit understanding that the theories thereby created shall never achieve their perfect, final form—that the conversation shall never terminate.

The discussion that follows, because of the complexity inherent in each side of the synthesis, has been divided into two parts. Part I consists of an examination of the ontological innovations and consequences of quantum mechanics; therein, the decoherence family of interpretations will be introduced and contrasted with a representative selection of other interpretations. Part II consists of the correlation of the ontological interpretation of quantum mechanics explored in Part I with Whitehead's cosmological scheme—both "mechanically" in terms of the phases of concrescence exemplified by quantum mechanical state evolution, and conceptually, in terms of Whitehead's nine Categoreal Obligations as fundamental principles presupposed and exemplified by the mechanics.

When considering this distinction between "mechanical" and "conceptual," however, one must take care to avoid conflating the concept of "mechanism" with the concept of "materialism"—a conflation that lies at the heart of the conventional connotation of "mechanism." Both quantum mechanics and Whiteheadian metaphysics describe a nondeterministic, nonmaterialistic process. But it is a mechanical process nonetheless, evinced in two aspects. First, it entails a realistic physics and metaphysics, grounded upon the objective actuality of the past; second, potentia are ontologically significant components of this process. They are integrated and reintegrated with other data into matrices of probability-valued subjective forms according to a set of governing principles (Whitehead's Categoreal Obligations, the various postulates of quantum mechanics, etc.)—principles capable of representation as rule-governed, mathematically describable constructions. Thus it is a nondeterministic, nonmaterialistic process that both Whiteheadian concrescence and quantum mechanical state evolution describe, both conceptually and mechanically; it is mechanism devoid of misplaced concreteness.

The intent of this book, then, is to suggest a narrow, phase-by-phase, concept-by-concept correlation of quantum mechanics and Whitehead's metaphysical scheme—that is, a correlation that avoids any omissions of the conceptual and comparative phases of Whitehead's "supplementary stage" of concrescence. Such omissions are often thought warranted when applying Whiteheadian philosophy to the physical sciences because of the presumed pertinence of these phases exclusively to conscious, high-order mental processes. This misplaced presumption might be due in part to Whitehead's choice of the term "mental pole" as an alternative to the "supplementary stage" of concrescence, which for some readers unfortunately implied a Cartesian connotation even though the repudiation of Cartesian mind-matter dualism is a fundamental principle of Whitehead's metaphysical scheme.

By contrast, other attempts to correlate Whiteheadian metaphysics with quantum mechanics (particularly the information theoretic interpretations) have tended to elevate the operations of the mental pole to primacy. In these syntheses, the spatiotemporal coordinative operations of the physical pole (the "primary stage" of concrescence) are often either merged into the supplementary stage/mental pole, or they are done away with altogether. The intent of such approaches seems to be to render the spatiotemporal extensiveness of actualities and systems of actualities, as well as any theory describing such extensiveness (such as Einstein's special and general theories of relativity), as mere abstractions derivable entirely from fundamental quantum events, in the same way that the concept of "material body" is so derivable in Whiteheadian metaphysics. By such interpretations, Whitehead's "fallacy of misplaced concreteness" finds its exemplification not only in the conventional notion of "fundamental materiality" but also in the conventional notion of "fundamental extensiveness in spacetime."

But for Whitehead, the spatiotemporally *extensive* morphological structure of actualities and nexūs of actualities given via "coordinate division" of actuality, primarily pertinent to the physical pole, is as crucial to concrescence as the *intensive* features of their relations given via "genetic division" of actualization, primarily pertinent to the mental pole. This close relationship, attended to in detail in chapter 5, is a key aspect of the dipolarity of concrescence in Whiteheadian metaphysics and the avoidance of a Cartesian bifurcated Nature.

It is hoped, then, that the close, concept-by-concept correlation proposed in this book will serve to demonstrate how quantum mechanics, as a fundamental physical exemplification of Whitehead's metaphysical scheme, might be heuristically useful toward a sound understanding of this scheme, and vice versa. Quantum mechanical concepts presented in Part I will thus be easily recognized when encountered in their analogous Whiteheadian forms in Part II; and likewise, readers already familiar with Whitehead will recognize these forms in their analogous quantum mechanical incarnations in Part I.

NOTES

1. Abner Shimony, "Quantum Physics and the Philosophy of White-head," *Boston Studies in the Philosophy of Science,* ed. R. Cohen and M. Wartofsky (New York: Humanities Press, 1965), 2:307.

2. Henry Folse, "The Copenhagen Interpretation of Quantum Theory and Whitehead's Philosophy of Organism," *Tulane Studies in Philosophy* 23 (1974): 32.

3. George Lucas, *The Rehabilitation of Whitehead: An Analytic and His-torical Assessment of Process Philosophy* (Albany: State University of New York Press, 1989), 189–199.

4. Alfred North Whitehead, *Process and Reality: An Essay in Cosmology, Corrected Edition,* ed. D. Griffin and D. Sherburne (New York: Free Press, 1978), 3.

5. Folse, "The Copenhagen Interpretation," 46–47.

6. See, for example, Whitehead, *Process and Reality,* 116.

7. David Bohm, "Time, the Implicate Order, and Pre-space," in *Physics and the Ultimate Significance of Time,* ed. David R. Griffin (Albany: State University of New York Press, 1986), 198.

8. Ibid., 196.

9. Ibid., 199.

10. David Bohm and B. J. Hiley, *The Undivided Universe: An Ontological Interpretation of Quantum Theory* (London: Routledge, 1993), 358.

11. Ibid., 361.

12. Whitehead, *Process and Reality,* 8.

13. Niels Bohr, *Atomic Theory and the Description of Nature* (Cambridge: Cambridge University Press, 1934).

14. Werner Heisenberg, *Physics and Philosophy* (New York: Harper and Row, 1958), 185.

15. Whitehead, *Process and Reality,* 6.

16. Wojciech Żurek, "Letters," *Physics Today* 46, no. 4 (1993): 84.

17. Ibid.

18. John von Neumann, *Mathematical Foundations of Quantum Mechan-ics* (Princeton, N.J.: Princeton University Press, 1955).

19. Whitehead, *Process and Reality,* 285.

20. Ibid., 40.

21. Ibid., 149.

22. Ibid., 149.

23. Ibid., 44.

24. Ibid., 43.

25. Ibid., 164.

26. Ibid., 285–286.

27. Ibid., 24.

28. Ibid., 65.

29. Ibid., 84.

30. Ibid., 84.

31. Ibid., 45.

32. Ibid., 29.

33. Ibid., 150.

34. Ibid., 228.

35. Ibid., 286.

36. Ibid., 116.

37. The most thorough and systematic example of an interpretation of quantum mechanics that explicitly recognizes these desiderata is to be found in Roland Omnès's *The Interpretation of Quantum Mechanics* (Princeton, N.J.: Princeton University Press, 1994); although some readers might find them too technical, his "21 Theses," which pertain to these philosophical desiderata, are extremely useful.

38. Whitehead, *Process and Reality*, 5.

39. Ibid., 278.

40. J. Monod, *Chance and Necessity: An Essay on the Natural Philosophy of Modern Biology* (New York: Vintage, 1972), 180.

41. James Hartle and Stephen Hawking, "The Wave Function of the Universe," *Physical Review D* 28 (1983): 2960–2975.

I

The Philosophical Implications
of Quantum Mechanics

2

The Ontological Interpretation of Quantum Mechanics

IN ANY MODERN EXPLORATION of the philosophy of science, one is at some point doomed to encounter that old, strangely constructed bridge that is quantum mechanics, the path across which seems to provide anything but sure footing. On one side of the chasm lie the conventional materialist ontologies, exemplified via the inductions of classical physics—inductions from our usual experiences of the macroscopic world. In general, these materialist ontologies characterize nature as fundamentally atomic and fluid—a collection of bits of matter whose motions, among other qualities and interactions, are deterministic and continuous through space and time. On the other side of the chasm lie modern experiences of the microscopic world, which, when initially accounted for by Planck and Einstein via the conventional materialist ontologies, characterized nature as a collection of material particles as before, but whose motions and interactions, when measured, appeared discontinuous and probabilistic rather than fluid and deterministic.

Bohr's 1913 model of the atom embodies this initial envisagement of the quantum mechanical microscopic world through the lens of mechanistic materialism. In this model, electrons are posited to be fundamentally material particles occupying a number of possible spatial "stationary states" around a fundamentally material nucleus. But rather than moving continuously through space from state to state according to previous conceptions, the electrons in Bohr's model must be thought of as making quantum leaps from one fixed state to another, each state associated with a discrete volume of space a certain distance from the nucleus and associated with a specific energy level. An electron making such a leap, in other words, must be thought of as making an instantaneous transition from one volume of space to another *without* moving through the space in between. Consideration of when or where an electron is during its transition from one state to another is rendered nonsensical in this model, despite the sensibility of such considerations given the materialist ontological framework Bohr's model otherwise requires. The demand for this troublesome caveat derives from the inescapable fact that any calculative prediction of an electron's state will yield only a probability as to which state it will occupy at the conclusion of a measurement, and never a unique factual outcome of the sort rendered by classical mechanics. And since the accuracy of these predicted probabilities is always confirmed retrodictively by experiment, the induction toward a "quantum" characterization of nature's most fundamental elements is, as concerns the science of physics, as justified as the materialist ontologies that preceded it.

The desire for a bridging of these two equally justifiable yet seemingly incompatible conceptions of nature—that of the conventional materialism of classical physics on one hand, and that inspired by the "old" quantum theory of Planck, Einstein, and Bohr on the other—fueled the subsequent work of Heisenberg, Bohr, Schrödinger, von Neumann, deBroglie, Dirac, Born, Jordan, and others during the years 1924–1928. But their efforts, while productive of a far more systematized and elegant quantum formalism, only served to emphasize and augment, rather than mitigate and reduce, the incompatibility of an ontological framework induced from classical mechanics and one induced from quantum mechanics. Granting for the moment that quantum mechanics is indeed ontologically significant at all, Abner Shimony distills these incompatibilities into five conceptual innovations implied by the quantum theory: (i) objective indefiniteness, (ii) objective chance, (iii) objective nonepistemic probability, (iv) objective entanglement, and (v) quantum nonlocality.[1]

Though the ontological implications of these innovations will be explored later in this chapter, it will be helpful to suggest now that there is a common source from which all five spring—a source that can be traced back even to the "old" quantum theory embodied by Bohr's atomic model: Though quantum mechanics makes use of facts—the facts constitutive of that which is being measured, for example, and the facts constitutive of the result of measurement—quantum mechanics cannot *account* for these facts, nor are the "mechanics" of quantum theory productive of facts. The mechanics are productive of probabilities only. When used to predict and therefore describe a measurement interaction, quantum mechanics makes use of facts stipulated to exist antecedent to the measurement interaction—facts that account for that which is to be measured, as well as the apparatus that will perform the measurement. These facts are, according to the quantum formalism describing the measurement interaction, causally productive of a matrix of mutually exclusive and exhaustive probabilities—probabilities referent to a unique, factual outcome. Although this unique outcome is anticipated by the mechanics, it is not accounted for by the mechanics, and therein lies the key to the conflict among various competing interpretations of the quantum theory. For many theorists, typically those most heavily invested in the classical mechanical worldview, the failure of quantum mechanics to account deterministically for a unique outcome is indicative of the incompleteness of the theory. Thus the five conceptual innovations suggested by Shimony are, for such theorists, properly viewed as epistemological artifacts to be cleared up either by augmenting the theoretical formalism or by improving experimental technology.

For other theorists, the failure of quantum mechanics to account for the unique outcome states it presupposes and anticipates is really no failure at all; it is instead properly recognized as a logical, and indeed, ontological limitation operative in any scientific theory. With respect to a conceptually analogous case suggested by Murray Gell-Mann,[2] when one wishes to predict the probability of a horse's winning a race, one necessarily presupposes (i) the fact of the horse (and everything else the race entails) antecedent to the race; and (ii) the fact of the horse's winning, or losing, at the conclusion of the race. By this view, even classical mechanics makes such presuppositions, which it cannot account for via its equations and formulas—namely, the very existence of the material particles whose motions and interactions are described by the mechanics.

That quantum mechanics is unable to account for its presupposed and anticipated facts has proven especially troublesome for some, given that quantum mechanics purports to characterize their relationship, via the Schrödinger equation, such that the matrix of probable measurement outcomes is not only subsequent to the measurement interaction and all the antecedent facts such an interaction presupposes, but also causally *consequent* of this interaction, which includes not only the facts constitutive of that which is mea-

sured, *but also the facts of the apparatus performing the measurement.*
John Bell states the difficulty thus:

> When it is said that something is "measured" it is difficult not to
> think of the result as referring to some preexisting property of the
> object in question. This is to disregard Bohr's insistence that in quan-
> tum phenomena the apparatus as well as the system is essentially in-
> volved. If it were not so, how could we understand, for example, that
> "measurement" of a component of "angular momentum" . . . in an
> arbitrarily chosen direction . . . yields one of a discrete set of values?[3]

The outcome of a quantum mechanical measurement, in other
words, is consequentially affected, at least to some degree, by that
which measures—a conclusion described by Heisenberg's famous
uncertainty relations. In their original 1927 incarnation, these relations were primarily intended to describe wave-particle duality with
respect to position and momentum measurements and the Comp-
ton Effect manifest in such measurements: Measurement of the position or velocity of a particle, according to Heisenberg, necessarily
entails the collision of the particle to be measured with particles
associated with the measuring apparatus. These collisions will always
affect position or velocity measurements such that, among other
consequences, a simultaneous position and velocity measurement
upon the same particle is impossible.

Given this particular application of the uncertainty relations, it
might appear that they are more epistemically significant than ontologically significant—that they have more to do with a deficiency in
the measuring procedure or apparatus than with the actualities being
measured. However, a more generalized incarnation of Heisenberg's
uncertainty relations in terms of the unavoidable change in a system's energy caused by a measurement interaction provides a more
fundamental and ontologically significant description: For example,
a measurement with duration Δt is made of the energy of a system,
and the measurement interaction causes an uncertainty or change
ΔE in the energy measured, the magnitude of which is given by:

$$\Delta E \Delta t \geq \frac{h}{4\pi}$$

where h is Planck's constant.

Thus, the *outcome* of a quantum mechanical measurement is

shown to be consequentially affected by that which measures. Unfortunately, however, this conclusion is often taken to imply that *that which is measured* via quantum mechanics is somehow causally influenced by that which measures—the source of the familiar lamentation that quantum mechanics has permitted the intrusion of sheer subjectivity into physics.

Such an alarming implication, however, should not be inferred from the Heisenberg uncertainty relations and the quantum formalism itself; it can only be inferred from the quantum formalism as interpreted via a classical materialist ontology, according to which a material object with factual material qualities antecedent to measurement continuously endures throughout measurement, thus remaining the same object at the conclusion of the measurement interaction—though by the Heisenberg uncertainty relations, with potentially different factual qualities somehow caused by the measurement. In other words, the object endures measurement by such an interpretation, yet its factual qualities change *because of* measurement and the particular apparatus used to perform the measurement. It is only when one attempts to interpret quantum mechanics via this classical materialist ontological notion of the qualification of an enduring substance by quality that quantum mechanics appears to thus threaten, via its supposed implication of sheer subjectivity, the objective reality of the world.

Alternatively, an ontological interpretation of quantum mechanics derived from the formalism itself, rather than from the formalism classically mediated, leads one toward a much different picture of the fundamental constituents of nature—again, granting for the moment that quantum mechanics is ontologically significant at all; one must keep in mind that all one can say for certain about quantum mechanics is that it is a theory by which physicists are able to successfully predict probabilities that specific types of measurements, under specific types of circumstances, will yield specific types of results. Whether or not this theoretical instrument is truly incapable of accommodation by the conventional materialist ontologies which otherwise reign supreme in physics is, to this day, a matter of heated debate, and arguments affirming a classical ontological interpretation of quantum theory have yet to be presented in this discussion.[4] That stated, one can consider, even at this very early stage of the discussion, the following chain of reasoning productive of innovative ontological generalizations from the quantum formalism:

1. Quantum mechanics describes the evolution of initial facts antecedent to a measurement interaction to final novel facts subsequent to a measurement interaction. It describes this evolution via the linear, deterministic Schrödinger equation, which yields a matrix of probable novel facts, from which a unique final fact will obtain indeterministically. The term "measurement interaction," therefore, is a relational term that presupposes the objective existence of the facts thus related by the quantum mechanical evolution. Without these facts, there is nothing to relate, and the concept of measurement becomes meaningless.

2. Quantum mechanics reveals, per the Heisenberg uncertainty relations, that the facts subsequent to a measurement interaction are also *consequent of* the measurement interaction, and in this sense, facts presuppose measurement as much as measurement presupposes facts. Bohr thus rightly induces from quantum mechanics "the impossibility of any sharp separation between the behaviour of atomic objects and the interaction with the measuring instruments which serve to define the conditions under which the phenomena appear."[5] According to quantum mechanics, then, the concepts of "fact" and "measurement" are mutually implicative; each is incapable of abstraction from the other, and both are therefore equally fundamental concepts. Put another way, facts are necessarily interrelated in quantum mechanics such that the notion of a single fact in isolation is meaningless.

3. Quantum mechanics therefore further reveals, per the Heisenberg uncertainty relations, that interrelations among facts are productive of novel facts, given that a measurement outcome is a novel fact (or an ensemble of novel facts) *consequent of,* and not merely subsequent to, the interrelations among facts comprising the measuring apparatus and the system measured. This is correctly interpreted not as a principle of sheer subjectivity, but as the fundamental physical exemplification of an ontological principle of relativity, such that any fact or system of facts cannot be considered apart from its interrelations with other facts. Measurement is an example of such interrelations.

Necessarily interrelated facts are therefore, according to the quantum formalism, the most fundamental constituents of nature capable of description via the physical sciences—and hence, by ontological induction, the most fundamental constituents of nature herself. Apart from such an induction, quantum mechanical "facts" are merely epistemically significant in their role as purely subjective qualifications of a material substance by quality. According to any ontology strictly induced from the quantum formalism, then, all the materialistic conceptions of nature—conceptions, for example, of material bodies whose motions, among other qualifications, are continuous through space and time—become abstractions from the most fundamental constituents of nature, which are discontinuous, yet necessarily interrelated, quantum facts. The notion, then, of a classically "isolated" object of measurement in quantum mechanics cannot be taken as ontologically significant, but merely as a conceptual abstraction; a fact, by its ontologically necessary interrelations with other facts as induced from the quantum formalism, can never be isolated from other facts in the classical sense.

The obvious implication is that every measurement interaction somehow involves the entire universe of all facts—an idea clearly alluded to by Heisenberg, for example, when he writes, "the transition from the 'possible' to the 'actual' takes place as soon as the interaction of the object with the measuring device, and thereby with the rest of the world, has come into play."[6] And indeed, a great many contemporary quantum theorists whose work will be explored in chapter 3 have demonstrated that the necessary interrelation among all facts in the universe is not only an interesting ontological induction one might draw from quantum mechanics; rather, theorists such as Robert Griffiths, Wojciech Zurek, Murray Gell-Mann, and Roland Omnès, among many others, have demonstrated that the interrelations between "measured" facts and facts belonging to the "environment" englobing the measurement interaction actually play a crucial mechanical role in the production of the measurement outcome.

BOHR'S STRATEGY: "COMPLEMENTARITY" AS EVIDENCE OF AN UNKNOWABLE OBJECTIVE REALITY

The conception of necessarily interrelated facts as nature's most fundamental elements constitutes the first major departure from conventional materialist ontologies that hold that measurement necessarily presupposes fact—"measurement" in this sense being a qualification of a material substance by quality—but fact need not presuppose measurement, such that a fact, as well as the material body it qualifies, can be considered in isolation. Bohr's insistence that even quantum mechanical measurements must be made with a classically described apparatus exemplifies the initial attempt to at least partially mediate quantum mechanics with a classical material-ist ontology in order to avoid this troublesome, mutually implicative relationship between fact and measurement. That this classical me-diation prescribed by Bohr utterly contradicts his insistence on "the impossibility of any sharp separation" between the system measured and the apparatus measuring, as exemplified by the Heisenberg un-certainty relations, has of course been the source of a great deal of confusion; the difficulty manifests itself in the formalism, for exam-ple, when one attempts to account for the correlations between alter-native potential microscopic measured-system outcomes and their respective alternative potential macroscopic measuring apparatus outcomes—a difficulty frequently referred to as "the problem of measurement" in quantum mechanics.

Indeed, most, if not all, of the conceptual difficulties and "para-doxes" associated with quantum mechanics can be traced to at-tempts to mediate or even wholly accommodate the latter within a classical mechanical ontology. There have been, over the years, a great number of proposed interpretations of quantum mechanics in-tended to solve these difficulties—some of which attempt to more coherently account for quantum mechanics as an abstraction from classical mechanics, and some of which attempt to account for classi-cal mechanics as an abstraction from quantum mechanics. Many of these ontological proposals will be explored later in this chapter, but it will be useful here to introduce, and then quickly dispense with, an interpretation that stands apart from all others: the instrumentalist interpretation—more a mindset than an interpretation, given that it denies that any interpretation of quantum mechanics is necessary. It is a tradition advocated, in the words of Popper, by physicists who have turned away from interpretations of quantum mechanics "be-cause they regard them, rightly, as philosophical, and because they believe, wrongly, that philosophical discussions are unimportant for physics. . . . [It is] a tradition which may easily lead to the end of science and its replacement by technology."[7]

This tradition is sometimes confused with Bohr's "pragmatic interpretation" of quantum mechanics, which is unfortunate given Bohr's obvious interest in the philosophical implications of the quantum theory. Though to confuse the two would be unfair, it would certainly be fair to say that Bohr's pragmatic interpretation significantly inspired the instrumentalist mindset popular among many physicists today; for Bohr prescribes, as we have seen thus far, two utterly incompatible conceptions of a quantum mechanical measurement interaction, for no other reason than it is practical to do so. On the one hand, he explicitly proscribes the possibility of any "sharp separation between the behaviour of atomic objects and the interaction with the measuring instruments which serve to define the conditions under which the phenomena appear."[8] Yet, on the other hand, Bohr prescribes treating the measuring apparatus classically, so that one is able to consider the facts pertaining to that which is measured as "external" to the apparatus, and therefore merely externally related, rather than mutually interrelated, thus preserving the conventional materialist notions of "subject" measuring "object." Likewise, the facts pertaining to the environment englobing that which is to be measured are similarly treated, such that the facts to be measured quantum mechanically can be considered as a "closed system," external to the surrounding environment.

An important justification for maintaining this classical conception of a closed, noninteracting system as the "object" of measurement is that the linear, deterministic Schrödinger equation at the very heart of quantum mechanics is applicable only to such closed systems. Bohr's prescription, however, also renders quantum mechanics nonuniversal and therefore ontologically insignificant, since the applicability of the mechanics becomes predicated upon an arbitrary dividing line separating the quantum and the classical.

That in itself may or may not be seen as a drawback, given that for practical purposes quantum mechanics is simply a tool used to predict the outcomes of measurements under specific conditions—not universal conditions. However, given Bohr's insistence that the interrelations between the apparatus and the atomic objects measured render them inseparable, as expressed via the Heisenberg uncertainty relations, this utterly contradictory classical, subject-object ontological mediation of quantum mechanics necessarily implies the following: The object of measurement, though classically isolated from the subject which measures, is nevertheless influenced somehow by this subject. Again, it is this conclusion that has led many to lament that quantum mechanics severely vitiates, via its implied sheer subjectivity, the objective reality of the world—a grievous and paradoxical violation of the same materialist characterization of quantum mechanical measurement which predicates such a conclusion. The ontologically classical characterization of quantum mechanical measurement, prescribed by Bohr, implies a relationship between subject and object that violates this very characterization.

One may attempt to escape such a conclusion, however, by appealing to the previous conclusion that quantum mechanics is non-universal, and therefore ontologically insignificant—at least to the extent that it requires no ontological innovation; its significance is epistemological only. The Heisenberg uncertainty relations can therefore be interpreted as holding that only one's *knowledge* of that which is measured—knowledge yielded via the outcome of a measurement—depends on the apparatus performing the measurement. The apparatus does not influence that which it measures, but merely affects one's knowledge of that which is measured. That which is to be known about an object, then, depends on the questions asked. This pragmatic characterization of quantum mechanical measurement as prescribed by Bohr is a component of what has come to be commonly referred to as the Copenhagen Interpretation of the quantum formalism, and although it perhaps serves to mitigate the perceived intrusion of subjectivity into physics, it does not remove this intrusion altogether. For according to the ontological requirements of classical physics—which the Copenhagen Interpretation attempts to subsume at least conceptually in its intended role as "*die endgültige Physik*" (the "absolutely final" physics)—the knowledge of an object, gleaned via measurement, presupposes (i) the objective existence of the object measured; and (ii) that the knowledge of such an object constitutes a factual qualification of the object—that "a fact of knowledge" is "knowledge of a fact." Therefore, if the factual qualification of an object is affected by the subject making this qualification, as implied by quantum mechanics, and a factual qualification of an object derives from real "qualities" of the object as required by the conventional materialism of classical physics, then one must conclude, based on these premises, that qualifications of an object not only derive from the qualities of an object, but also *affect* the qualities of an object.

Therefore, even the characterization of quantum mechanics as merely epistemically significant does not prevent a violation of the classical, materialist ontology from which the premises of measurement—as prescribed by Bohr—derive. Even if one were to retreat to the notion that knowledge of an object gleaned via measurement does not necessarily reveal factual qualities pertaining to the object—that "a fact of knowledge" yielded by a quantum mechanical measurement does not necessarily entail "knowledge of a fact"—the objectively real material world thus immunized from sheer subjectivity is immunized only by the belief that the actual facts of such a world are essentially unknowable at all. Yet it is precisely this retreat which Bohr advocates as the best way to bridge quantum and classical mechanics: "In physics," he writes, "our problem consists in the co-ordination of our experience of the external world,"[9] such that "in our description of nature the purpose is not to disclose the real essence of phenomena but only to track down as far as possible relations between the multifold aspects of our experience."[10]

Thus, where the physical sciences were conventionally held to reveal factual qualities of nature, the objective reality of which was thought to be demonstrable by the apparently universal laws inducible from these qualities, Bohr suggests that quantum mechanics necessarily leads one to a much different conception of physics: The laws of physics, once thought to reveal the essence of nature's most fundamental constituents, now merely reveal subjective coordinations of our experiences of nature. "In quantum mechanics," Bohr writes, "we are not dealing with an arbitrary renunciation of a more detailed analysis of atomic phenomena, but with a recognition that such an analysis is *in principle* excluded" (emphasis in original).[11] Physical laws, for Bohr, must be seen as qualifications of experience only, and the apparent regularity of those experiences capable of description by a physical law cannot be taken as evidence for or against a qualitative characterization of nature herself. For if one were to do so, one would be confronted with the fact that such inductions from the laws of quantum mechanics on one hand, and from the laws of classical mechanics on the other, each lead to entirely incompatible characterizations of nature. Better to say that the laws of quantum mechanics and classical mechanics reveal incompatible yet "complementary"—and for some unknown reason, exceedingly regular—coordinations of our experiences of nature. It is by means of such complementary coordinations of experience that one may bridge quantum and classical mechanics as mutually applicable to the task of characterizing our experiences of nature without characterizing nature herself.

Bohr's proscription against the induction of a fundamental characterization of nature from the laws of quantum mechanics—or classical mechanics—is the cornerstone of his proposed bridge linking quantum and classical mechanics; and in laying this cornerstone, he of course violates that very proscription. For it is clear that it was indeed quantum mechanics that led Bohr to impose the epistemic sanctions embodied by his "principle of complementarity"—and those sanctions clearly constitute an *ontological characterization of nature* whose essence, by this very ontology, consists of a fundamentally *unknowable* reality,[12] subjectively experienced at two levels: (i) publicly, when these subjective experiences are regular enough to be described via physical laws which appear to hold universally; (ii) privately, when these subjective experiences are not regular enough to be described via a physical laws. These ontological generalizations, which Bohr clearly induced from quantum mechanics, and which clearly violate their own desiderata by the qualification of an "unknowable reality" as "unknowable," closely resemble, according to Henry Stapp, the ontology proposed by William James,[13] which characterizes the world as consisting of (i) "hypersensible realities"; (ii) public "sense objects" (those subjective experiences whose coordination is regular enough to be expressible by physical laws); (iii) "private concepts" (those experiences that are truly "subjective" by the conventional meaning of the term, and incapable of coordination expressible by physical laws).[14]

The central claim of the Copenhagen Interpretation is that quantum mechanics is complete—that it cannot be abstracted from a more fundamental characterization of reality. Bohr's justification for this claim via the Jamesian-style ontology he prescribes, however, hinges on the even broader claim that reality is incapable of objective characterization at all. Since it is only one's experiences of reality that can be characterized and coordinated, the seeming incompatibility between the ontological implications and presuppositions of classical mechanics and those of quantum mechanics is resolved by stripping both of any ontological significance at all. Quantum and classical mechanics are thus relegated to the level of merely epistemically significant complementary coordinations of experience, and as such their incompatibility becomes unimportant. The practical advantages of this Jamesian ontological accommodation of the Copenhagen Interpretation are, then, as follows:

First, it allows for the classical conceptions of "subject" and "object" as *externally related*, such that a measurement outcome reveals the state of the object subsequent to the measurement interaction. One may, then, arbitrarily divide the components of a quantum mechanical measurement interaction into an object of measurement, classically isolated from the subject apparatus performing the measurement, as well as from the external environment englobing both. The evolution of the classically closed object/system through the measurement interaction is then capable of description via the linear, deterministic Schrödinger equation, which can be applied only to such a classically closed system.

Second, it allows for the nonclassical conceptions of subject and object as *internally related*—per the Heisenberg uncertainty relations—such that a measurement outcome reveals the state of both subject and object as consequent of, not merely subsequent to, the measurement. In this way, subject and object are nonclassically inseparable, since quantum mechanics describes the effect of the subject apparatus upon the measurement outcome, even when the apparatus and the object measured are spatially well separated.

These allowances were, for Bohr, thought to be possible only by virtue of his proviso requiring an ultimately unknowable reality, such that these wholly incompatible conceptions of subject and object in measurement do *not* constitute incompatible characterizations of the "real things" measuring and measured, which would render quantum mechanics incoherent and incomplete; they are instead to be considered only as complementary coordinations of our experiences of these things. Apart from this Jamesian ontological accommodation of the Copenhagen Interpretation, then, Bohr's "principle of complementarity" would amount to little more than an argument for the inescapable ontological incoherence of quantum mechanics, and therefore the completeness of quantum mechanics—the central

claim of the Copenhagen Interpretation—would be anything but demonstrable.

Von Neumann's Alternative to Bohr's Epistemic Schism: Objective Reality via a Coherent Ontological Interpretation

The two advantages of the Jamesian ontology prescribed by Bohr are,
however, readily achievable without adopting this conflicted ontology and its notion that reality is most fundamentally hypersensible
and incapable of revelation via our experiences. And indeed, the
completeness of the Copenhagen Interpretation—the notion that it
entails the most fundamental characterization of our experiences of
nature—is quite sensible apart from such an ontology. For as discussed at the beginning of this chapter, the most fundamental constituents of nature one might induce from quantum mechanics are
necessarily interrelated facts from which all other conceptions—
including classical notions of material objects with continuous motions and other attributes—can be shown to be abstractions. The
ontological significance of the classical notion of an "objective reality" is preserved in the sense that interrelated facts are *facts*; measurement, as an example of such interrelation, presupposes the facts
to be related—the fact of that which is measured, the fact of that
which performs the measurement, and the anticipated fact of the
outcome of the measurement. And the ontological significance of
the Heisenberg uncertainty relations—wherein facts and measurement are mutually implicative—is guaranteed via the conceptually
innovative notion of *necessarily interrelated* facts, such that the existence of a fact cannot be considered in isolation from other facts. In
other words, since the fundamental constituents of nature are not
just facts, but necessarily interrelated facts, fact presupposes measurement (interrelations with other facts) as much as measurement
presupposes fact. The creation of a novel fact in quantum mechanics, in the form of a measurement outcome, requires and is conditioned by its interrelations with the facts antecedent to it—facts
pertaining to that which is measured (system), but also to that which
measures (apparatus) and, by implication, even to that which is not
measured (environment).

According to such an ontological framework, an "experience of reality," as Bohr used the term, is ultimately the ontologically fundamental interrelation among facts—facts pertaining to we who experience, interrelated with the facts of the world we experience. The concept of "experience," then, is by this ontology identical to the concept of "measurement"; both are exemplifications of the ontologically fundamental notion of necessarily interrelated facts. An "experience of physical reality," given Bohr's use of the term, thus becomes a chain of interrelations: The interrelations of facts constituting that which is measured with the facts constituting the measuring apparatus; and the interrelation of facts constituting the apparatus with facts constituting the experimenter who observes the apparatus, et cetera. Bohr's concept of a hypersensible reality is no longer necessary, since *experience* itself—necessarily interrelated facts—is ontologically fundamental, rather than an underlying, essentially unknowable material "object" subjectively qualified by experience. The notion of quantum mechanical measurement as a chain of interrelated facts constituting that which is measured and that which measures, however, implies that the measuring apparatus—envisioned as a classical object by Bohr and thus in some sense isolated from the quantum mechanical system that it measures—should instead be envisioned fundamentally as an ensemble of facts interrelated quantum mechanically with the facts being measured.

This was the approach proposed by von Neumann[15] as a way of mathematically accounting for the correlations between the facts constituting the measuring apparatus and the facts constituting that which is measured. Moreover, if quantum mechanics is to be a truly coherent, universal theory, von Neumann suggests that these correlations should further extend to the facts constituting that which "measures" the measuring apparatus—in other words, the body and mind of the human observer. In this way, the classical conceptions of "subject of measurement" and "object of measurement" become properly understood as arbitrary abstractions from a more fundamental quantum mechanical characterization of measurement as the correlation of serially ordered quantum actualizations. Every particular subject-object correlation, then, becomes a datum for a subsequent subject-object correlation. It is only by such a scheme of "psycho-physical parallelism," suggests von Neumann, that the innovative and classically problematic "subjective" features of quantum mechanics might be mediated with the necessary "objective" realism in which modern science is grounded.

It is a fundamental requirement of the scientific viewpoint—the so-called principle of the psycho-physical parallelism—that it must be possible so to describe the extra-physical process of the subjective perception as if it were in reality in the physical world—i.e., to assign to its parts equivalent physical processes in the objective environment, in ordinary space. (Of course, in this correlating procedure there arises the frequent necessity of localizing some of these processes at points which lie within the portion of space occupied by our own bodies. But this does not alter the fact of their belonging to the "world about us," the objective environment referred to above.) In a simple example, these concepts might be applied about as follows: We wish to measure a temperature. If we want, we can pursue this process numerically until we have the temperature of the environment of the mercury container of the thermometer, and then say: this temperature is measured by the thermometer. But we can carry the calculation further, and from the properties of the mercury, which can be explained in kinetic and molecular terms, we can calculate its heating, expansion, and the resultant length of the mercury column, and then say: this length is seen by the observer. Going still further, and taking the light source into consideration, we could find out the reflection of the light quanta on the opaque mercury column, and the path of the remaining light quanta into the eye of the observer, their refraction in the eye lens, and the formation of an image on the retina, and then we would say: this image is registered by the retina of the observer. And were our physiological knowledge more precise than it is today, we could go still further, tracing the chemical reactions which produce the impression of this image on the retina, in the optic nerve tract and in the brain, and then in the end say: these chemical changes of his brain cells are perceived by the observer. But in any case, no matter how far we calculate—to the mercury vessel, to the scale of the thermometer, to the retina, or into the brain, at some time we must say: and this is perceived by the observer. That is, we must always divide the world into two parts, the one being the observed system, the other the observer. In the former, we can follow up all physical processes (in principle at least) arbitrarily precisely. In the latter, this is meaningless. The boundary between the two is arbitrary to a very large extent. In particular we saw in the four different possibilities in the example above, that the observer in this sense needs not to become identified with the body of the actual observer: In one instance in the above example, we included even the thermometer in it, while in another instance, even the eyes and optic nerve tract were not included. That this boundary can be pushed arbitrarily deeply into the interior of the body of the actual observer is the content of the principle of the psycho-physical parallelism—but this does not change the fact that in each method of description the boundary must be put somewhere, if the method is not to proceed vacuously, i.e., if a comparison with experiment is to be possible. Indeed experience only makes statements of this type: an observer has made a certain (subjective) observation; and never any like this: a physical quantity has a certain value.[16]

One of the implications of this interpretation of quantum mechanical measurement is that the unique qualities of a particular actual "subject" in some way govern the correlations between that subject and its particular object. In this way, the set of probability outcomes yielded by quantum mechanics for any given measurement in a chain of measurements, such as the chain described above, "fits" the particular qualities of *that* subject within the chain. Von Neumann noted that the quantum mechanical mechanism governing these subject-object correlations is very different from the quantum mechanical mechanism by which a *unique* measurement outcome is actualized for any given measurement interaction. For one thing, the mechanism governing subject-object correlations does not yield a unique measurement outcome, but rather a mixture of probable measurement outcomes. This and other distinguishing features led von Neumann to further suggest that a coherent, universal interpretation of quantum mechanics requires that the process of subject-object correlation in a quantum mechanical measurement interaction must be distinct from the process descriptive of a unitary wavefunction evolution in such an interaction. He thus proposes a "Process 1" productive of subject-object correlation and the influence of such correlation on the evolution of the *mixture* of probable measurement outcomes yielded by quantum mechanics; and "Process 2" describes the causal, unitary wavefunction evolution to a *particular* probable outcome state. Process 1, in other words, is a non-unitary and thus "non-causal" evolution explicative of the subject-object correlation characteristic of the particular mixture of probable measurement outcomes yielded by this process; this particular mixture of probable outcome states, in other words, would have been otherwise *if* the subject had been otherwise. Process 2, by contrast, is descriptive of a more generic, unitary, and thus "causal" evolution of the wavefunction to a particular probable outcome state—an evolution von Neumann characterizes as an "automatic" change, as opposed to the "arbitrary" change effected by a specific subject-object measurement interaction described by Process 1. Since quantum mechanics cannot account for the existence of actualities (though it can describe their evolution), Process 2 is merely descriptive of the unitary evolution from initial factual system state to final probable system state. But Process 1 is explicative, in that it accounts for the particular probability outcomes yielded; they are functions of the necessary quantum mechanical correlations between a particular subject (i.e., a particular measuring apparatus) and a particular measured system. Von Neumann writes:

> Why then do we need the special Process 1 for the measurement? The reason is this: In the measurement we cannot observe the system S by itself, but must rather investigate the system $S + M$, in order to obtain (numerically) its interaction with the measuring apparatus M. The theory of the measurement is a statement concerning $S + M$, and should describe how the state of S is related to certain properties of the state of M (namely, the positions of a certain pointer, since the observer reads these). Moreover, it is rather arbitrary whether or not one includes the observer in M, and replaces the relation between the S state and the pointer positions in M by the relations of this state and the chemical changes in the observer's eye or even in his brain (i.e., to that which he has "seen" or "perceived"). . . . In any case, therefore, the application of [Process] 2 is of importance only for $S + M$. Of course, we must show that this gives the same result for S as the direct application of [Process] 1 on S. If this is successful, then we have achieved a unified way of looking at the physical world on a quantum mechanical basis.[17]

Von Neumann's "Process 1" and its relevance to the modern decoherence-based interpretations of quantum mechanics will be explored further in chapter 3. For now, let us return to his more general thesis that an ontologically coherent interpretation of quantum mechanics requires that both measuring apparatus and measured system be treated quantum mechanically. Though von Neumann's proposal would seem warranted by an ontology of fundamentally interrelated facts, the overwhelming complexity of a quantum mechanically described macroscopic measuring apparatus would entail calculations far too unwieldy, if not impossible, to be employed in practice. This criticism, however, cannot be taken as a sensible argument against

his proposal; for the notion of necessarily interrelated facts as ontologically fundamental was induced *from* quantum mechanics by virtue of the fact that the formalism presupposes this notion. As
mentioned earlier, quantum mechanics does not account for the existence of facts, nor can it, since a quantum mechanical description
of a measurement interaction presupposes the existence of facts. By
the same token, quantum mechanics does not *explain* the interrelation of facts; it presupposes this interrelation, and though it *describes*
this interrelation, it cannot *account* for it. To fault quantum mechanics for its inability to explain the existence of necessarily interrelated
facts presupposed by the mechanics is as unreasonable as to fault
classical mechanics for its inability to explain the existence of matter, similarly presupposed. The notion of matter as ontologically fundamental was an induction made from classical mechanics and its
description of matter; and the notion of necessarily interrelated facts
as ontologically fundamental is an induction made from quantum
mechanics and its description of these facts.

The conceptual advantage of von Neumann's approach is that it
exemplifies the fundamental ontological concept that the process by
which facts are interrelated should pertain to *all* facts—not just
those facts which are arbitrarily "isolated," constituting that which
is measured. The Schrödinger equation, which describes this interrelation, should apply to the facts comprised by the measuring apparatus as well as to the facts comprised by that which is measured.
And, by implication, the facts comprised by the person observing (or
"measuring" or "experiencing") the measuring apparatus should, by
their interrelations with the facts of the apparatus and the system
measured, also evolve according to the Schrödinger equation during
the measurement interaction. Again, if the Schrödinger equation
were used to describe the evolution of such a so-called von Neumann
chain of interrelated facts, the calculations required would be entirely unmanageable; but conceptually, such a chain of interrelated
facts is quite reasonable given the ontological inductions made thus
far—and quite necessary if these inductions are to be coherent and
consistent.

Von Neumann's program exemplifies the requirement, for the sake
of ontological coherence, that the quantum mechanical interrelation
of facts be universal, such that when we treat the measuring apparatus as a classical object as physicists do in practice, it is explicitly
understood that the classicality of the apparatus is not an ontological

characterization, but merely a conceptual abstraction from its more fundamental description as an ensemble of interrelated facts, which are themselves interrelated with the facts pertaining to that which is measured. So that when it seems as though the apparatus as a classical object, separated from that which it measures, somehow affects that which it measures, one is able to dispense with the abstraction and reclaim the underlying, ontologically fundamental notion of mutually and necessarily interrelated facts as the ultimate constituents of nature.

CLOSED SYSTEMS

We should recall, however, that treating the measuring apparatus as a classical object—whether as an abstraction from a more fundamental ontological conception, or as a "complementary" way of subjectively coordinating our experiences of a more fundamental "hypersensible reality"—is not merely to make the Schrödinger equation manageable; it is to make the Schrödinger equation applicable at all. For as mentioned earlier, the Schrödinger equation can be applied only to a classically "closed" system, and it is the treatment of the measuring apparatus as a classical object, isolated from that which it measures, which effectively renders the system measured "closed." It is, in this sense, difficult to suppose how an ontology based on mutually, necessarily interrelated facts can accommodate the requirement that *some* facts—those constituting the arbitrarily defined closed system—are incapable of interrelations with other facts. And since the Schrödinger equation lies at the heart of quantum mechanics, from which we have drawn our ontological inductions thus far, the coherence of these inductions depends upon their accommodation of this equation. For it is this equation that describes the mechanics of the interrelations among facts in a measurement interaction, and most significantly, it qualifies these relations as causally efficacious; without this qualification, then, quantum mechanics would be rendered utterly incapable of describing our experiences of physical causality.

The means by which a closed system can be abstracted from ontologically fundamental interrelated facts is quite simple, however, and is no threat at all to the coherence of this ontology; in fact, it is the requirement that the ontology *be* coherent and universal which provides the solution, already alluded to by von Neumann: If reality fundamentally consists of necessarily and mutually interrelated facts, then the only closed system capable of accommodation by such an ontology is, of course, the system of *all* facts. This would mean that the interrelations of *some* facts in a measurement interaction somehow entail the interrelations of all facts—that is, the entire universe. It would require, in other words, that even facts which are neither measured nor measuring—facts belonging to the "environment" englobing the measurement at hand—play some role in the quantum mechanical measurement of selected facts. Every quantum mechanical measurement, then, somehow must be thought to involve the entire universe—an implication, as indicated before, stressed by advocates of the decoherence-based interpretations of quantum mechanics that will be explored in chapter 3.

But even von Neumann's proposal to treat the measuring apparatus quantum mechanically as an ensemble of interrelated facts was sufficient to render the calculations unmanageable. It therefore seems inconceivable that the experimental environment—let alone the entire universe—must be accounted for quantum mechanically as facts interrelated with the facts of the system measured and the measuring apparatus—particularly since the environment, unlike the measuring apparatus, seems to play no appreciable role in the orthodox quantum formalism. Although the universal interrelation of all facts is entirely justified as an implication of the ontological inductions made from quantum mechanics thus far, it is natural that those physicists who would embrace such inductions would want to see this implication exemplified functionally somehow by quantum mechanics itself. And indeed, as physicists, Griffiths, Omnès, Gell-Mann, Zurek, and their like-minded colleagues stress the *physical* necessity of this implication—the quantum mechanical *function* of the interrelation between "environmental" facts and facts comprised by the system measured and the measuring apparatus—more than they do the philosophical necessity. And for them, environmental facts not only play a role in quantum mechanics, but a crucial role.

At this stage of the discussion, however, it will be helpful to emphasize that this notion of the universe as the only "closed system" involves an equally crucial conceptual innovation—one that is the hallmark of those interpretations of quantum mechanics which embrace this concept of "closed system": The "environment," via its necessary interrelations with all other facts, is thus able to "make measurements" as well as any apparatus or human observer, and indeed does so continuously given the necessary interrelations of all facts within the closed system of the universe. Whereas "measurement" was typically characterized either tacitly or explicitly anthropically as the interrelations among facts comprised by a human-observed "measuring apparatus" and facts comprised by a "measured system," the necessary interrelations of all facts within a closed system not only bring the facts comprised by the environment into play; the "measurement" of one "object" (subsystem of facts) by an "apparatus" (another subsystem of facts) in any given experimental procedure is now seen as a specific exemplification of the mutual interrelations among all facts, including those belonging to the environment, in the wider closed system.

This, of course, provides welcome relief to those who would lament that quantum mechanics entails the intrusion of necessary human subjectivity into physics; "facts as *consequent of* measurement" in quantum mechanics implied, for some, that facts would not exist at all unless measured by an observer—that the moon would exist with precise factual qualities such as position, size, shape, only when observed, or that a cat in a closed box containing a vial of poison that may or may not have spilled is somehow neither "factually alive" nor "factually dead" until the box is opened and the cat is observed. When the universe is properly considered as a closed system, however, one is able to acknowledge the interrelations between the cat and moon subsystems with facts environmental to these subsystems: Photons from the universal microwave background radiation are scattered off the moon, thus continuously "measuring" it; the molecules of the box interact with the cat as well as with the outside world, so the cat is similarly in a constant state of "measurement." Whereas a classical ontological conception of the cat, or any other item in the universe, characterizes it as an object which endures measurement (or the lack thereof), the ontological conception suggested by quantum mechanics characterizes the cat as a subsystem of facts in a state of perpetual creation by their interrelations with all other facts in the universe—a universe which, by the same process, is therefore itself in a perpetual state of creation. The classical conception of material existence thus becomes an abstraction from a more fundamental conception of creative factual interrelations.

In the closed system of the universe, all facts are interrelated, and therefore the interrelations of some facts—the measurement of one subsystem by another—necessarily entails interrelations with facts environmental to these subsystems. This "environmental monitoring" of measured systems in quantum mechanics is, as mentioned before, more than conceptually significant as regards the ontological coherence it affords to quantum mechanics; it is also directly and pragmatically reflected in the mathematical formalism of the mechanics, from which further interesting ontological inductions and hypothetical deductions such as those suggested in the paragraph above can be made, and that will be explored later in chapter 3.

Leaving aside, for the moment, the formal quantum mechanical function of the interrelations between environmental facts and facts belonging to the system measured and the measuring apparatus, it is clear that consideration of the universe as the only truly closed system affords quantum mechanics a genuine ontological coherence and consistency only approximated or vaguely suggested by other interpretations. It is a coherence and a consistency far more substantial, for example, than that supposedly provided by Bohr's principle of complementarity, with its arbitrary dividing line demarking nature's quantum and classical essences as we experience and coordinate them via quantum and classical mechanics, respectively. And yet it must be emphasized that quantum mechanics does not merely entail the mutual interrelations among all facts comprised by the universe as the singular "closed system"; rather, quantum mechanics entails the interrelation of all facts antecedent to a *given measurement interaction* with facts subsequent to and, by the Heisenberg uncertainty relations, consequent of, this *particular* measurement interaction. Quantum mechanics, then, describes the interrelations among facts comprised by a closed system—the universe—such that these interrelations are always (i) productive of a consequential novel fact (the measurement outcome); and therefore (ii) always *relative to that consequential novel fact*, such that the novel fact in production is consequentially related with all facts antecedent to it.

The Ontological Significance of Potentia

Since, however, it is the *production* of a consequent novel fact—that is, a measurement outcome—that is conditioned by its interrelations with antecedent facts, and not the factual outcome itself (for as a "fact," a measurement outcome is already settled, "objectively real," and cannot be conditioned, influenced, altered, or undone), an ontology based on mutually related facts requires an additional innovation—the concept of a *potential* fact. Potential facts as mathematical components of the quantum formalism provide the means by which the creation of a novel fact, such as a measurement outcome, can be causally related with antecedent facts in quantum mechanics.

Heisenberg, in his analysis of the ontological significance of quantum mechanics, insists upon the fundamental reality and function of potentia in this regard. For him, potentia are not merely epistemic, statistical approximations of an underlying veiled reality of predetermined facts; potentia are, rather, ontologically fundamental constituents of nature. They are *things* "standing in the middle between the idea of an event and the actual event, a strange kind of physical reality just in the middle between possibility and reality."[18] Elsewhere, Heisenberg writes that the correct interpretation of quantum mechanics requires that one consider the concept of "probability as a new kind of 'objective' physical reality. This probability concept is closely related to the concept of natural philosophy of the ancients such as Aristotle; it is, to a certain extent, a transformation of the old 'potentia' concept from a qualitative to a quantitative idea."[19]

Quantum mechanics thus becomes characterized as the mechanics of interrelations among facts—among "actualities"—toward the production of novel actualities (i.e., measurement outcomes), and these interrelations are mediated by potentia—the real "things" upon which the mechanics operate. Therefore, a coherent ontology induced from quantum mechanics as suggested by Heisenberg in the preceding quotations presupposes not one but two fundamental constituents of nature—two species of reality: (i) necessarily interrelated facts, or "actualities"; (ii) potential facts, or "potentia," which provide the means by which a novel fact is causally interrelated with the facts antecedent to it. One would expect that if potentia are, indeed, one of the two fundamental species of reality implied by quantum mechanics, that potentia and their ontological function would find some exemplification in the quantum formalism. And they do, in the form of the matrix of probabilities yielded in a quantum mechanical measurement interaction. These probabilities, terminal of every quantum mechanical measurement interaction, are understood to be potential outcomes—potential facts which will be constitutive of the measured system after measurement.

When one speaks of the "state" of a system, one is referring to a maximal specification of these facts, so that in quantum mechanics one is concerned with (i) the initial, actual state of the system prior to the measurement interaction; (ii) the matrix of mutually exclusive and exhaustive potential outcome states predicted by quantum mechanics, each of whose propensity for actualization is valuated as a probability; and (iii) the actual, unique outcome state observed at the end of the measurement. It is always the case that only a single outcome state is ever observed—that only one potential outcome state ever becomes actual in any given measurement interaction. But it is also the case that the "actualization" of this unique potential outcome lies beyond the scope of quantum mechanics, which yields only a matrix of probable outcomes and never a singular, determined outcome. This difficulty is typically referred to as the "problem of state reduction" or the "problem of the actualization of potentialities."

One might connote in these two terms the idea of a physical, dynamical reduction mechanism which should be thought to operate upon the matrices of potentia yielded by a quantum mechanical measurement interaction. For if potentia are indeed ontologically significant—that is, "real"—there should be, it has been suggested, some equally "real" physical mechanism describing their dynamical evolution to a unique, actual state, in the same way that the Schrödinger equation describes the dynamical evolution from the initial, unique, actual state to the matrix of potential outcome states. Quantum mechanics, by such an argument, is therefore incomplete and in need of augmentation. Some physicists have developed a variety of proposals to this end—stochastic and nonlinear modifications of the Schrödinger equation intended to account mechanically for the actualization of potentia. The ontological implications of some of these programs will be explored presently, but for now it is enough to note that these proposed remedies for the incompleteness of quantum mechanics were inspired by the belief in the ontological significance of potentia as suggested by Heisenberg. The difficulty is that these proposals treat potentia as though they were actualities, conflating the two concepts such that the Schrödinger equation is interpreted as producing "coexistent" *actual* alternative measure-

ment outcomes that must, then, be physically and dynamically reduced to a single outcome.

The conceptual difficulties produced by such conflation of actuality and potentiality in quantum mechanics were most notably fore-
warned by Schrödinger himself in his infamous "cat" hypothetical,
conveyed in a brief paragraph of his 1935 essay "Die gegenwartige
Situation in der Quantenmechanik."[20] A cat, placed in a box, is sub-
jected to a procedure wherein it will either live or die as a result. A
quantum mechanical measurement of the cat, then, yields a matrix
of two probability outcomes. The problem of state reduction manifests here in two questions: (i) If quantum mechanics is ontologically
significant, then the matrix of two probability outcomes yielded (cat-
alive and cat-dead) must be ontologically significant; if this is so,
then why do we never experience the monstrosity of a live-dead cat
superposition terminal of a quantum mechanical measurement? (ii)
Since a quantum mechanical measurement of the cat yields such a
matrix or superposition of coexisting states, yet observation always
yields a unique state wherein the cat is always either dead or alive
and never both, what is the physical mechanism by which the superposition or matrix of coexistent, "real" states is reduced to a unique
"real" state, and when is this mechanism effected? Many physicists
have concluded that, as mentioned earlier, the Schrödinger equation
requires modification to account for this elusive mechanism; for unless one were inclined to assign the same ontological significance to
"potentia" as is assigned to "actualities" in classical mechanics, one
must interpret the matrix of probable outcome states yielded by
quantum mechanics as a matrix of coexisting actual states.

The conception of potentia as a separate species of reality—
ontologically, rather than merely epistemologically significant—has
not been widely embraced by physicists biased toward a classical ontological accommodation of quantum mechanics. And indeed, most
of the proposed solutions to the problem of state reduction entail
attempts to fit quantum mechanics into a classical ontological
framework. Many of these attempts have proven to be extremely
unappealing because they either entail presuppositions even more
radical than the concept of "real" potentia suggested by Heisenberg
and championed by Popper, or because they have been experimentally disconfirmed. They are, nevertheless, important to any exploration of the ontological significance of quantum mechanics, for they

have helped to disclose the standards of coherence, logical consistency, empirical applicability, and empirical adequacy sought after in its interpretation.

The particular classical, materialist ontological interpretations of quantum mechanics referred to above characterize the matrix of probable measurement outcomes as a set of coexistent, actualized (classically real) alternatives, the superposition of which is dynamically reduced by some physical process to yield a unique actuality. The proposal of Eugene Wigner is representative of this approach to state reduction. For Wigner, the continuous evolution of the state of the measured system as described by the Schrödinger equation is inconsistent with the seemingly discontinuous actualization of a unique outcome state. He suggested that this inconsistency might be remedied by a nonlinear modification of the Schrödinger equation, which would account for a dynamical mechanism by which the matrix of alternative real states is discontinuously reduced to a single state. Wigner further suggested that the mechanism described by this nonlinear modification might even be attributed to the influence of the mind of the observer upon that which is measured, which would account for why we never observe superpositions or matrices of alternative states in nature; for the act of observation itself *is* the nonlinear mechanism causative of state reduction.[21] One finds a similar conclusion given by Walter Heitler, who wrote: "One may ask if it is sufficient to carry out a measurement by a self-registering apparatus or whether the presence of an observer is required." Heitler concluded that "the observer appears, as a necessary part of the whole structure, and in his full capacity as a conscious being."[22]

It should be stressed that despite occasional assertions to the contrary, Heisenberg did not ultimately hold this view as an appropriate induction from quantum mechanics. For Heisenberg, again, potentia are ontologically significant constituents of nature that provide the means by which the facts comprising the measuring apparatus and the facts comprising the system measured (as well as those comprising the environment) are interrelated in quantum mechanics: "The transition from the 'possible' to the 'actual' takes place," he writes, "as soon as the interaction of the object with the measuring device, and thereby with the rest of the world, has come into play; it is not connected with the act of registration of the result by the mind of the observer."[23] The conflation of Heisenberg's ontological inductions and those of Wigner, Heitler, and others is, however, under-

standable; for Heisenberg's ontologically innovative contribution to
the Copenhagen Interpretation was amalgamated with the severe
epistemic sanctioning of Bohr's worldview. Heisenberg thus inter-
prets the matrix of probable outcome states not as an objective integration of potentia—despite his belief that potentia are "real" in the
Aristotelian sense—but rather as a mathematical representation of
our incomplete knowledge of the evolution of the measured system.
Though he clearly holds that, ontologically, the integration of potentia is productive of, and will "reduce to," a unique actuality, the
density matrix itself is, for Heisenberg, merely an epistemic reflection of these potentia. And the discontinuous reduction of the density matrix is, likewise, an epistemic reflection of an increase in one's
knowledge of the facts that have actualized. Heisenberg explains that
although the ontological actualization of potentia is not driven by
any physical mechanism associated with the conscious registration
of the actualized by the mind of the observer, "the discontinuous
change in the probability function, however, takes place with the act
of registration, because it is the discontinuous change of our knowledge in the instant of registration that has its image in the discontinuous change of the probability function."[24] Heisenberg's concept of
potentia thus seems to imply both an unavoidable subjectivity veil-
ing nature and, at the same time, her reliable objective reality. Patrick Heelan explains this dual function thus:

> The objective tendency or *potentia* . . . is on the one hand *not* simply
> the thing-in-itself in the external world, *nor* on the other hand is it
> simply the transcendental ego; it bridges both the external world and
> the transcendental subjectivity of the knower. As Heisenberg wrote in
> the *Martin Heidegger Festschrift* (1959), "the search for the natural
> laws of the [ultimate structure of matter,] entails the use of general
> principles of which it is not clear whether they apply to the empirical
> behaviour of the world or to *a priori* forms of our thought, or to the
> way in which we speak."[25]

Most of the confusion with respect to Heisenberg's concept of potentia and its dual implication of subjectivity and objectivity in nature has to do with the habitual classical tendency to apply the
fundamental mediative function of potentia solely to facts belonging
to (i) "measured system" and (ii) "measuring apparatus"—that is,
the classical "object" and "subject"; the result is that the broader,

more accurate role of potentia is overlooked. For a coherent ontological interpretation of quantum mechanics reveals that most fundamentally, potentia are mediative of *facts*, whether they belong to the measured system, the measuring apparatus, the human observer, or the universal environment—for all are subsumed by the singular closed system of the universe. It is not the subjective interaction of "human" facts with object "measured system" facts that somehow alters or actualizes the latter; it is, rather, the interrelations among all facts relative to a given fact or subsystem of facts, which are productive of potentia which will evolve to become novel facts. With their most fundamental role in mind, then, it is clear that Heisenberg's potentia are not indicative of a fundamental subjectivity pervading nature, but rather a fundamental relativity.

For many, Heisenberg's epistemological interpretation of the density matrix has overshadowed and obfuscated the ontological implications he proposed—the reality of potentia and the idea that novel actualities are somehow produced by a process involving the integration and valuation of these potentia. The conflation of his interpretation of state reduction and those of Wigner, Heitler, and others, is an unfortunate consequence of this obfuscation, but understandable in light of it. Much theoretical work has been done in recent years, however—proposals by several physicists mentioned earlier, such as Roland Omnès, Robert Griffiths, Murray Gell-Mann, and Wojciech Żurek, to name a few—which begins with Heisenberg's ontological interpretation of potentia. These theories then depict, explicitly, the evolution of the state of a measured system—the integration and valuation of the potentia associated with this system—as an ontological process, rather than merely an epistemological representation of an underlying ontology, as given by Heisenberg. These interpretations will be explored more fully in the following chapter, but they are mentioned here to suggest that the ontological characterization of state reduction implied by Wigner and Heitler, wherein the density matrix is depicted as an integration of coexisting actualities reduced by some physical mechanism to a unique actuality, is not the only ontological characterization that physicists have considered. For the depiction of the density matrix as an integration of coexisting actualities, though it would certainly render quantum mechanics ontologically significant, exacts a price many physicists are unwilling to pay—the implication that until the box containing Schrödinger's cat is opened, the cat exists as a superposition of two alternative, yet equally actual states.

Yet for many physicists, this price is entirely reasonable given the undeniable value of an ontological interpretation of quantum mechanics. The "Continuous Spontaneous Localization" theory of GianCarlo Ghirardi, Alberto Rimini, and Tulio Weber (GRW), based on early work by Phillip Pearle in the 1970s, also treats the superposition of states yielded by the Schrödinger equation as a superposition of actual physical states, upon which there should operate some physical mechanism causative of reduction to a unique state such that macroscopic superpositions like Schrödinger's cat are extremely short-lived. Like Wigner, GRW propose a modification of the Schrödinger equation to this end; but whereas Wigner proposed a characterization of state reduction as ultimately discontinuous via a nonlinear modification, the spontaneous localization theory characterizes state reduction as ultimately continuous, via a linear modification. The proposed mechanism for unitary reduction represented by this modification is a stochastically fluctuating field, which continuously and spontaneously "collapses" the superposition of states into a unique state.[26] Where particle density is high, as in the case of a macroscopic object such as a cat, these stochastic field fluctuations cause extremely rapid collapses such that, according to John Bell, "any embarrassing macroscopic ambiguity in the usual theory is only momentary in the GRW theory. The cat is not both dead and alive for more than a split second."[27] However, the disconcerting ontological implications of the cat's being simultaneously alive and dead at all, even if for only a split second, cannot be overlooked. Among other difficulties is the implied double-violation of the logical principles of non-contradiction and the excluded middle— violations which undermine one of the first principles of modern science: the presupposed correlation of causal relation and logical implication.

And indeed, the spontaneous localization theory of GRW is intended to be ontologically significant given that it purports to specify "the physical reality of *what exists out there*"[28] according to Ghirardi. The spontaneous localization theory is to provide, he writes, "a mathematically precise formalism allowing a unified description of all phenomena, containing a single fundamental dynamical principle that governs all processes."[29] Thus, the theory not only implies the coexistence of alternative versions of entities like Schrödinger's cat, even if only for a split second; even more troubling, it entails an ontology whose "fundamental dynamical principle" is the continuous destruction of most of these entities. Indeed, it would seem that any interpretation of quantum mechanics that treats the density matrix as comprising, against Heisenberg's admonition, coexisting, alternative actualities rather than coexisting potentia, would require an ontological principle whereby the vast majority of these entities are destroyed—whether continuously, per GRW's proposal, or discontinuously, per Wigner's proposal.

The "Relative State" interpretation of Hugh Everett—often referred to as the "many worlds interpretation"[30]—purports to avoid this troubling ontological implication. It is similar in its treatment of the density matrix as an amalgam of coexisting, actual alternative states; but whereas the interpretations discussed earlier proposed the need for a physical mechanism to reduce this matrix to a unique state, Everett suggests that no such mechanism or modification of the Schrödinger equation is necessary, given that the concept of a single unique outcome state is itself unnecessary. We never experience a superposition of equally real alternative states because each alternative state can be thought of as a unique outcome state occurring in its own relative universe. An alternative state is, in this way, always relative to its particular universe in the same way that a temporal duration, by the theory of special relativity, is always relative to its particular inertial frame.

A key advantage of Everett's proposal is that it provides a conceptual account of the correlations between the facts comprised by the measuring apparatus and those comprised by the system measured; like von Neumann, Everett suggests that both the apparatus and the system measured evolve quantum mechanically. But whereas von Neumann was unable to account for the correlations between the separate evolutions of the system and apparatus, Everett is able to do so by virtue of the fact that each probable system outcome, and its correlated apparatus outcome, occurs in the same "relative universe." One universe contains a live cat, correlated with an observer who sees a live cat, and another universe contains a dead cat, correlated with an observer who sees a dead cat.

Though Everett's proposal avoids the implicit ontological principle by which most coexistent, alternative entities must be destroyed, there remains the difficulty of accepting an ontology that entails that there be multiple "copies" of oneself—an ontology where there is no distinction between potentiality and actuality. Such an implication is arguably far more radical and philosophically problematic than that suggested by Heisenberg, wherein potentia are fundamentally "real," though not "actual," constituents of nature. Indeed, an ontology where everything that can happen does happen renders meaningless the idea of things that *should* happen. "If such a theory were taken seriously," writes John Bell, "it would hardly be possible to take anything else seriously. So much for the social implications."[31]

Common to all these proposals and a great many others is the conflation of actuality and potentiality symptomatic of any attempt at a classical accommodation of quantum mechanics. The "reality" of the alternative probability outcomes yielded by quantum mechanics is indeed acknowledged in these theories, but it is reality as classically defined—reality monopolized by actuality—such that Schrödinger's cat in its bizarre live-dead superposition is necessarily as real as any other cat. Heisenberg's conception of potentia as ontologically significant, fundamental constituents of nature answers this mischaracterization, such that the matrix of alternative states yielded by quantum mechanics is a matrix of coexisting *potential*— but nevertheless *real*—states, not a matrix of coexisting actual states. For Heisenberg, the concept of potentia, then, constitutes, in his words,

> a first definition concerning the ontology of quantum theory. . . . One sees at once that this use of the word "state," especially the term "coexistent state," is so different from the usual materialistic ontology that one may doubt whether one is using a convenient terminology. . . . One may even simply replace the term "state" by the term "potentiality"—then the concept of "coexistent potentialities" is quite plausible, since one potentiality may involve or overlap other potentialities.[32]

The problem of state reduction thus becomes redefined: The question is no longer, "What is the mechanism by which a unique actuality physically evolves from a matrix of coexistent actualities?" but rather, "What is the mechanism by which a unique actuality evolves from a matrix of coexistent potentia?"

The answer to that question can be found in the quantum formalism itself: The matrix of states yielded by quantum mechanics is not merely a matrix of potential states; they are, rather, mutually exclusive and exhaustive *probable* states. That is to say, they are a *selection* of potential states that have evolved, via the Schrödinger equation, to become probabilities—that is, potentia qualified by a valuation between 0 and 1, such that together, these probable states represent mutually exclusive and exhaustive potential outcomes, satisfying the logical principles of non-contradiction and the excluded middle, respectively. Thus one (and only one) outcome *must* occur. Unlike purely potential outcomes, then, probability outcomes clearly presuppose and anticipate the necessary actualization of a unique outcome; for without such an outcome, the concept of a probability itself is utterly meaningless. In this sense, the actualization of potentia is less a physical, dynamical function of quantum mechanics than it is a conceptual function logically presupposed by the mechanics.

As stated earlier, to attempt to account for the existence of facts via a mechanism that presupposes their existence is logically untenable. Nevertheless, the inability of quantum mechanics to account for a *unitary* evolution from a pure-state superposition to a unique measurement outcome—rather than to a matrix of alternative probable outcomes—is considered by many physicists to be a defect of the formalism, or as we have seen in the examples above, at least a problem that should be solved. Against such objections, we may simply recall the previously cited analogy suggested by Murray Gell-Mann: When one wishes to predict the probability of a horse winning a race, one necessarily presupposes (i) the fact of the horse (and everything else the race entails) antecedent to the race; and (ii) the fact of the horse's winning, or losing, at the conclusion of the race. In the same way, quantum mechanics cannot account for the existence of facts, given that (i) quantum mechanics presupposes their existence antecedent to the measurement interaction—facts which account for that which is to be measured, as well as the apparatus which will perform the measurement; and (ii) quantum mechanics *anticipates* their existence, via the yielded mutually exclusive and exhaustive probability outcomes, subsequent to and consequent of the measurement interaction.

Put simply, quantum mechanics does not describe the actualization of potentia; it only describes the *valuation* of potentia. The answer to the problem of state reduction is thus at once as simple and as elusive as the "problem" of the three interior angles of a triangle adding up to 180 degrees, or the "problem" of the existence of the universe in classical mechanics. In the case of the latter, most cosmologists have found it much more interesting to simply logically stipulate the existence of the primordial $t = 0$ initial conditions of the universe, in whatever form it may have had, and focus instead on the mechanics describing how those initial conditions have evolved to become what the universe is today, and thus glean what it might become tomorrow. Similarly, in quantum mechanics, the "problem of the actualization of potentia," or more generally, the problem of the existence of facts, is far less interesting than the mechanics that describe how facts are causally productive of potentia, and how these potentia evolve to become valuated probabilities subsequently and consequently anticipative—and somehow creative—of novel facts.

NOTES

1. A. Shimony, "Search for a Worldview Which Can Accommodate Our Knowledge of Microphysics," in *Philosophical Consequences of Quantum Theory: Reflections on Bell's Theorem*, ed. J. Cushing and E. McMullin (Notre Dame, Ind.: Notre Dame University Press, 1989), 27.

2. Murray Gell-Mann, *The Quark and the Jaguar: Adventures in the Simple and the Complex* (New York: W. H. Freeman, 1994), 141.

3. John Bell, "Against Measurement," in *Sixty-Two Years of Uncertainty: Historical, Philosophical, and Physical Inquiries into the Foundations of Quantum Mechanics*, ed. Arthur I. Miller (New York: Plenum, 1990).

4. Quantum mechanics is often interpreted as an incomplete classical theory, for example, where probability, nonlocality, and entanglement are ontologically insignificant epistemic artifacts. These artifacts are consequential of our inability to account for a continuum of classically deterministic "hidden variables," which would otherwise complete the theory. Max Born, in this spirit, demonstrated how quantum mechanical probability outcomes can be equated with classical statistical probability outcomes. This purely classical, statistical interpretation of quantum mechanics has, with respect to nonlocality, since been theoretically disconfirmed by John Bell (J. S. Bell, "On the Einstein Podolsky Rosen Paradox," *Physics* 1, no. 3 [1964]: 195–200), and later experimentally disconfirmed (A. Aspect, J. Dalibard, and G. Roger, *Phys. Rev. Lett* 44 [1982]: 1804–1807). Various "mostly classical" nonlocal hidden-variables interpretations, such as the one of David Bohm, became newly championed in response to these disconfirmations. Thus the relegation of quantum mechanics to mere epistemic significance by Bohm's interpretation, in defense of classicality, now

depends upon the concept of an etherlike field of point-particles causally influencing each other nonlocally, in flagrant violation of the otherwise classical ontology the interpretation was designed to preserve.

5. Niels Bohr, "Discussion with Einstein on Epistemological Problems in Atomic Physics," in *Albert Einstein: Philosopher-Scientist*, ed. Paul Arthur Schilpp (New York: Harper, 1959), 210.

6. Werner Heisenberg, *Physics and Philosophy* (New York: Harper Torchbooks, 1958), 54–55.

7. K. Popper, *Quantum Theory and the Schism in Physics* (New Jersey: Rowman and Littlefield, 1956), 100.

8. Bohr, "Discussion with Einstein," 210.

9. Niels Bohr, *Atomic Theory and the Description of Nature* (Cambridge: Cambridge University Press, 1934), 1.

10. Niels Bohr, *Atomic Physics and Human Knowledge* (New York: Wiley, 1958), 18.

11. Bohr, *Atomic Theory and the Description of Nature*, 1.

12. To posit an ontological characterization of the world as fundamentally unknowable is, of course, to doom the ontology to incoherence by its own reasoning, reducing it to the paradox of Epimenides. An admonition from Whitehead comes to mind: "The requirement of coherence is the great preservative of rationalistic sanity. But the validity of its criticism is not always admitted. If we consider philosophical controversies, we shall find that disputants tend to require coherence from their adversaries, and to grant dispensations to themselves" (*Process and Reality*, 6).

13. Henry P. Stapp, *Mind, Matter, and Quantum Mechanics* (Berlin: Springer-Verlag, 1993), 60.

14. William James, *Pragmatism and the Meaning of Truth* (Cambridge, Mass.: Harvard University Press, 1978), 239.

15. John von Neumann, *Mathematical Foundations of Quantum Mechanics* (Princeton, N.J.: Princeton University Press, 1955).

16. Ibid., 418–420.

17. Ibid., 352.

18. Heisenberg, *Physics and Philosophy*, 41.

19. Werner Heisenberg, "The Development of the Interpretation of the Quantum Theory," in *Niels Bohr and the Development of Physics*, ed. Wolfgang Pauli (New York: McGraw-Hill, 1955), 12.

20. Erwin Schrödinger, "Die gegenwartige Situation in der Quantenmechanik," *Naturwissenschaftern* 23 (1935): 807–812, 823–828, 844–849. English translation, John D. Trimmer, *Proceedings of the American Philosophical Society* 124 (1980): 323–338.

21. E. P. Wigner, in *The Scientist Speculates: An Anthology of Partly-Baked Ideas*, ed. Irving John Good (New York: Basic Books, 1962), 284.

22. Walter Heitler, "The Departure from Classical Thought in Modern Physics," in *Albert Einstein: Philosopher-Scientist*, ed. Paul Arthur Schilpp (New York: Harper, 1959), 194.

23. Heisenberg, *Physics and Philosophy*, 55.

24. Ibid.

25. Patrick Heelan, S.J., *Quantum Mechanics and Objectivity* (The Hague: Martinus Nijhoff, 1965), 151.

26. G. Ghirardi, A. Rimini, and P. Pearle, in *Sixty-Two Years of Uncertainty*, ed. A. Miller (New York: Plenum, 1990), 167, 193.

27. John Bell, *Speakable and Unspeakable in Quantum Mechanics* (Cambridge: Cambridge University Press, 1987).

28. G. Ghirardi, in *Structures and Norms in Science*, ed. M. L. Dalla Chiara et al. (Dordrecht: Kluwer Academic, 1997).

29. G. Ghirardi, *Physics Today* (Letters) April 1993: 15.

30. Hugh Everett III, *Rev. Mod. Phys.* 29 (1957): 44. See also B. S. DeWitt and N. Graham, eds., *The Many Worlds Interpretation of Quantum Mechanics* (Princeton, N.J.: Princeton University Press, 1973).

31. Bell, *Speakable and Unspeakable in Quantum Mechanics*, 159–168.

32. Heisenberg, *Physics and Philosophy*, 185.

3

The Evolution of
Actuality to Probability

IF, AS HEISENBERG SUGGESTED, potentia indeed constitute an ontologically fundamental species of reality, then the mathematical representation of the quantum mechanical evolution of the matrix of
potentia to the matrix of probabilities—that is, the valuation of potentia as probabilities between zero and one—ought to be heuristically useful in understanding the ontological implications of this
evolution. The mathematical representation should prove helpful in
visualizing not only the process described by quantum mechanics
but also, as we shall see in the following chapters, the process described by Whitehead. The basic formalism introduced and explored
in this chapter is intended to be comprehensible to readers unfamil-
iar with mathematics. Some familiarity with the Pythagorean theorem and the addition of vectors will be useful, but not necessary.

THE FORMAL DESCRIPTION OF QUANTUM
MECHANICAL STATE EVOLUTION

As mentioned earlier, we refer to the "state" of a system as a maximal specification of the facts or actualities comprised by the system.
(These facts/actualities are typically referred to by physicists as "observables" or "collective observables," even though most are, for all
practical purposes, unobservable; for this reason, we will use the
terms "facts" and "actualities" instead.) For the sake of simplicity,
let us consider an idealized system consisting of nothing other than
an old-fashioned traffic signal—the type with two lights, red and
green. Let us further suppose that the signal always functions nor-
mally, such that neither of the lights is burned out and that both
will never be illuminated simultaneously. And finally, let us ignore
everything other than the status of the two lights—the wiring, the

casing, and so on—such that the state of our idealized traffic-light system entails but one fact: The status of the signal, which is either green or red.

The state of any system in quantum mechanics is represented by a vector of unit length in Hilbert space—an abstract linear vector space that, in our idealized example, can be depicted via a simple x–y Cartesian coordinate system. The benefit of using Hilbert spaces in quantum mechanics is that these spaces are capable of representing, in a mathematically useful way, potentia as well as actualities and their relationship in a given system. This representation is based on two simple principles:

(i) Every physical system can be represented by a unique Hilbert space $\mathcal{H}_s$, and the state S of a given physical system (again, "state" being the maximal specification of the facts/actualities/observables associated with the system) can be represented by a single vector $|u\rangle$ of unit length in the system's Hilbert space.

(ii) In a measurement interaction involving the system, there is a one-to-one correspondence between the number of probability-valuated outcome system states and the number of dimensions comprised by the Hilbert space.

In the case of our highly idealized traffic-light system, which has two potential states S, then, the associated Hilbert space has only two dimensions: One represents "S = green," and the other "S = red." This particular two-dimensional Hilbert space is therefore easily represented by simple x–y coordinate axes, where a vector of unit length $|u_{green}\rangle$ along the x-axis represents "S = green" and a vector of unit length $|u_{red}\rangle$ along the y-axis represents "S = red."

As regards our idealized system, one can at this point see the logical need for Principle 2: It guarantees, in satisfaction of the logical principles of non-contradiction and the excluded middle, that there exist *some* potential states that are mutually exclusive and exhaustive, as represented by the mutual orthogonality of the x and y dimensions (where "mutually orthogonal vectors" are vectors at right angles to one another). To generalize this somewhat, we can say that in a Hilbert space of n dimensions, there are only n mutually orthogonal vectors, representing only n mutually exclusive states. If, for example, we were dealing with a more modern traffic-signal system with a green, a red, and a yellow light, our Hilbert space would require three dimensions, represented by x, y, and z coordinate axes. It

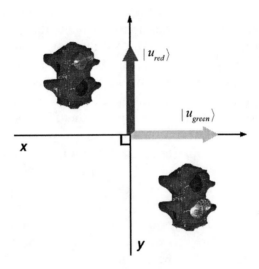

FIGURE 3.1 The two potential states S of a traffic signal represented by orthogonal vectors of unit length in a Hilbert space of two dimensions. The vector $|u_{green}\rangle$ along the x-axis represents "$S = green$," and the vector $|u_{red}\rangle$ along the y-axis represents "$S = red$."

is only by means of this third dimension that we are guaranteed three mutually exclusive and exhaustive states: $S = green$, $S = yellow$, or $S = red$, represented by the mutual orthogonality of the vectors along the x, y, and z axes. An added benefit of mutually orthogonal vectors representing mutually exclusive and exhaustive states is that such vectors can be grouped so that, as regards a modern traffic-signal system, we can specify the three possible states thus: S is either green, represented by $|u_{green}\rangle$, or S is not green, represented by the plane formed by $|u_{red}\rangle + |u_{yellow}\rangle$; S is either red $|u_{red}\rangle$ or not red $|u_{green}\rangle + |u_{yellow}\rangle$; S is either yellow $|u_{yellow}\rangle$ or not yellow $|u_{green}\rangle + |u_{red}\rangle$.

Such groupings are referred to as "subspaces" of the Hilbert space, and their usefulness becomes readily apparent when considering nonidealized systems with manifold—even innumerable—potential states, each with a multiplicity of associated facts/actualities/observables. Consider, for example, the state of "System *You*" as you read this chapter. As is the case for most physical systems, a maximal specification of the facts/actualities/observables associated with you is far too unwieldy to calculate, so let us focus, as is typically done in quantum mechanics, on just one fact/actuality/observable: the location of System *You* in the universe. We first represent the state of

System *You* via a vector of unit length somewhere in a Hilbert space of a certain number of dimensions—as many dimensions as there are places in the universe where you might be located. This is, of course, a practically infinite number of dimensions, regardless of our choice of location units—impossible to graph as we did with our idealized two- and three-dimensional examples. But conceptually, the principles are the same. We can say, then, that all of the practically infinite number of mutually orthogonal vectors of unit length in our Hilbert space represent potential, mutually exclusive and exhaustive states of System *You*. A certain number of these states include, among other facts/actualities/observables, locations of System *You* that are in the Amazon River. Again, the mutual orthogonality of the vectors provides for mutually exclusive potential states that, as regards location, prevent you from being two places at once. It is convenient to group all of these mutually orthogonal "Amazon River vectors" into a subspace $\mathscr{E}_{amazon}$.

Each vector represents a *maximal specification* of System *You*, including many facts/actualities/observables other than location; and whatever we might say about these other facts/actualities/observables, so long as a given vector belongs to the subspace $\mathscr{E}_{amazon}$, the location of System *You* is indeed in the Amazon. One vector in this subspace, $|u_{hot}\rangle$, for example, might represent a potential state where you are wearing a T-shirt in the Amazon, and another vector in this subspace, $|u_{cold}\rangle$, might represent a potential state where you are wearing a winter coat in the Amazon. These differences and a myriad of others aside, all vectors in the subspace $\mathscr{E}_{amazon}$ represent alternative states that share a single defining characteristic: the fact/actuality/observable "location" = "the Amazon River." The state of System *You* as regards location, then, can be expressed thus: you are either somewhere in the Amazon, or you are not. If the former, then your state is represented by some vector belonging to $\mathscr{E}_{amazon}$; if the latter, then your state is represented by some vector outside of $\mathscr{E}_{amazon}$, which is itself a subspace, $\mathscr{E}_{amazon}^{\perp}$ consisting of all vectors orthogonal to those in $\mathscr{E}_{amazon}$.

In this way—unlike the classical mechanical conception of measurement as a qualification along a fluid continuum of qualities (consider the image of a "fluidly" sweeping second hand on an electric analog clock face, for example)—the conception of measurement in quantum mechanics entails the discrete specification of the truth or falsity of a particular fact/actuality/observable in a potential system state. A digital clock either reads "12:23 P.M." or it does not. A laser's wavelength is 650 nm when measured, or it is not. If the former, the vector representing the laser's state belongs to $\mathscr{E}_{650\,nm}$ and if the latter, it belongs to $\mathscr{E}_{650\,nm}{}^{\perp}$. If a traffic light is green, its state vector belongs to $\mathscr{E}_{green}$, and if it is not green, its state vector belongs to $\mathscr{E}_{green}{}^{\perp}$.

This conception of measurement via explicit contrast—the discrete specification of the state of a system in terms of the truth or falsity (that is, the actuality) of a given potential fact associated with a potential state—and the mathematical representation of such contrasts via vectors in Hilbert space, provides not only a coherent framework for performing the relevant calculations of quantum mechanics, but also a coherent framework for exploring the relationship between potentiality and actuality in quantum mechanics, and the conceptually innovative ontological implications of this relationship.

The first step of this exploration begins with the following (and for many, infamous) quantum mechanical innovation, already introduced with a previous reference to Schrödinger's cat: There are vectors of unit length—that is, vectors that represent real states of a physical system—that belong *neither* to $\mathscr{E}$ nor to $\mathscr{E}^{\perp}$ for a given fact. There are, in other words, real physical states wherein the actuality of a given potential fact is *objectively indefinite*. Let us return to our old-fashioned, green-red traffic-light system, where a unit-length vector $|u_{green}\rangle$ along the x-axis represents the state $S = green$, and a unit-length vector $|u_{red}\rangle$ along the y-axis represents $S = red$. What of the unit-length vector $|\psi\rangle$, which lies neither on the x nor y axis? Since it is a vector of unit length, it represents a real potential physical state according to the quantum theory; but it represents a physical state wherein the traffic signal is neither green nor red—or similarly, a state wherein Schrödinger's cat is neither alive nor dead. Quantum mechanics describes the evolution of such "pure states" $|\psi\rangle$, whose potential facts are entangled in a superposition, into a finite set of mutually exclusive and exhaustive states (represented by a set of mutually orthogonal vectors). Each of these potential outcome states is assigned a probability, and all of the alternative outcome states together are referred to collectively as a "mixed state," so named because unlike a pure state $|\psi\rangle$, the mixed state comprises a mixture of valuated, mutually exclusive probable outcome states. In the case of our idealized, old-fashioned, red-green traffic-signal example, the pure state $|\psi\rangle$ evolves to become an *eigenstate* (one of the component states of the mixed state) where the signal is definitely either green $|u_{green}\rangle$ or red $|u_{red}\rangle$ upon observation (measurement). The state begins as the pure state, $|\psi\rangle$, where the two possible outcome states are correlated, or entangled. This is expressed as the sum of the associated vectors: $|\psi\rangle = |u_{green}\rangle + |u_{red}\rangle$. The vector $|u_{green}\rangle$ is added to $|u_{red}\rangle$ by simply transporting the tail of $|u_{green}\rangle$ to the head of $|u_{red}\rangle$ and then drawing a new vector $|\psi\rangle$ from the origin to the transported head of $|u_{green}\rangle$.

It is thus easy to see why the pure state vector $|\psi\rangle$ can be expressed as the sum of two eigenvectors $|u_{green}\rangle$ or $|u_{red}\rangle$. The probability assigned to each eigenstate of the mixture is indicated by adding a complex coefficient α and β to $|u_{green}\rangle$ and $|u_{red}\rangle$ respectively, which indicates the vector's length:

$$|\psi\rangle = \alpha|u_{green}\rangle + \beta|u_{red}\rangle$$

where $|\alpha|^2 + |\beta|^2 = 1$.

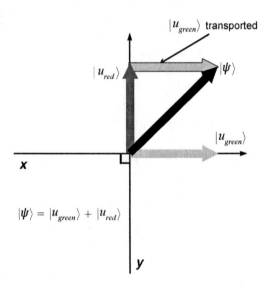

FIGURE 3.2 $|\psi\rangle$ is a vector of unit length, and therefore a real, but objectively indefinite state that is neither $|u_{green}\rangle$ nor $|u_{red}\rangle$ but is expressible as a sum of these two vectors. $|u_{green}\rangle$ is added to $|u_{red}\rangle$ by transporting the tail of $|u_{green}\rangle$ so that it falls upon the head of $|u_{red}\rangle$. The sum $|\psi\rangle$ is a vector drawn from the origin to the transported head of $|u_{green}\rangle$.

These coefficients express the probabilities of the alternative out-come states in the following way: α and β are "projections" of the vector $|\psi\rangle$ upon the eigenvector $|u_{green}\rangle$ and the eigenvector $|u_{red}\rangle$ respectively (see figure 3.4). So, α is the length of the projection of $|\psi\rangle$ upon $|u_{green}\rangle$, and this length represents the probability of $|u_{green}\rangle$ becoming actual. The projection of $|\psi\rangle$ upon $|u_{green}\rangle$ is written as:

$$\alpha = \|Pu_{green}\psi\|$$

Similarly, β is the length of the projection of $|\psi\rangle$ upon $|u_{red}\rangle$, written as:

$$\beta = \|Pu_{red}\psi\|$$

As one can see by looking at figure 3.4, $|\alpha|^2 + |\beta|^2 = 1$ is simply the necessary satisfaction of the Pythagorean theorem by the projec-tors; and as these coefficients are intended to express probabilities,

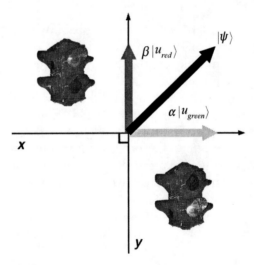

$$|\psi\rangle = \alpha|u_{green}\rangle + \beta|u_{red}\rangle$$

FIGURE 3.3. The pure state $|\psi\rangle$ as an objectively real, but objectively in-definite superposition of states $|u_{green}\rangle$ and $|u_{red}\rangle$, expressed as the sum of $|u_{green}\rangle$ and $|u_{red}\rangle$, with the complex coefficients satisfying $|\alpha|^2 + |\beta|^2 = 1$. This sum to unity of the probable outcome states represented by $|u_{green}\rangle$ and $|u_{red}\rangle$ thus accords with standard probability theory; $\alpha|u_{green}\rangle$ and $\beta|u_{red}\rangle$ are probability-valuated alternative outcome states.

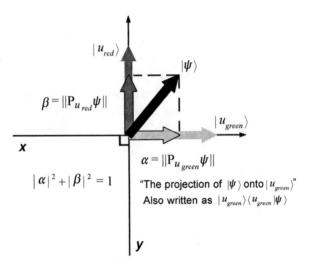

FIGURE 3.4. The complex coefficient α is the length of the projection of $|\psi\rangle$ upon $|u_{green}\rangle$, and the complex coefficient β is the length of the projection of $|\psi\rangle$ upon $|u_{red}\rangle$.

$|\alpha|^2 + |\beta|^2 = 1$ also satisfies standard probability theory. It is significant, however, that in quantum mechanics, neither $\alpha|u_{green}\rangle$ nor $\beta|u_{red}\rangle$ are vectors of unit length, such that neither represents a system state by itself; only when added together, correlated, do they sum to unity, thus representing a valid state. This is the infamous "Superposition Principle."

This correlated, pure state $|\psi\rangle = \alpha|u_{green}\rangle + \beta|u_{red}\rangle$ evolves to become a mixed state where the correlated superposition of eigenstates $\alpha|u_{green}\rangle + \beta|u_{red}\rangle$ is now expressible as a "density matrix" ρ of separate, uncorrelated mutually exclusive and exhaustive outcomes, each valuated as a probability.

As discussed earlier, quantum mechanics does not describe the evolution of the mixed state, to a unique state $|u_{green}\rangle$ or $|u_{red}\rangle$; quantum mechanics describes only the evolution of the much broader "pure state" superposition $|\psi\rangle = \alpha|u_{green}\rangle + \beta|u_{red}\rangle$ to the mixed state. Put another way, quantum mechanics does not include a mechanism for the actualization of potentia; it merely describes the *valuation* of potentia (via the complex coefficients α and β)—the valuation of the alternative potential eigenstates belonging to the mixed state, such that these alternative potential states become probabilities, and not just potentia.

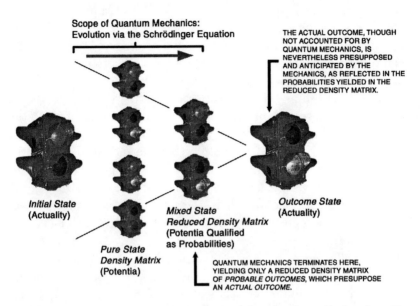

FIGURE 3.5. The quantum mechanical evolution of an idealized traffic-signal system.

The first, most obvious ontological issue raised by this mathematical, linear vector description of quantum mechanics concerns the expression of the initial state vector $|\psi\rangle$ as a sum of the potential outcome state eigenvectors. Even in our idealized traffic-light example, $|\psi\rangle$ can clearly be expressed as the sum of a great many orthogonal vectors other than $|u_{green}\rangle$ and $|u_{red}\rangle$. Why does the initial state vector $|\psi\rangle$ always evolve to become $|u_{green}\rangle$ or $|u_{red}\rangle$, rather than some other set of orthogonal vectors, representing other mutually exclusive and exhaustive outcome states, such as, for example, $|u_{purple}\rangle$ or $|u_{hazel}\rangle$? Why, in other words, is Schrödinger's cat necessarily a live or dead *cat* when the box is opened, and not a mouse or a dog?

From a strictly instrumentalist interpretation of quantum mechanics, this somewhat arbitrary selection of alternative mutually exclusive and exhaustive outcome states and their associated mutually orthogonal state vectors (referred to as the "preferred basis" or "pointer basis") poses no great problem, inasmuch as the alternative potential outcome states are always known prior to measurement. In other words, we simply *know* our traffic signal will either be red or green upon measurement, and not purple, hazel, aquamarine, and so on. As we will see later in this chapter, however, the ontological significance of the preferred basis really derives from the fact that a quantum mechanical system state always evolves *relative* to a particular quantum fact (the "indexical eventuality") associated with the apparatus performing the measurement. The preferred basis, in other words, is a feature of this apparatus and its contribution to the evolution of the system state. The ineluctability of this relativity implies, as von Neumann suggested, that the measuring apparatus *together with* the system measured should be interpreted as a composite quantum system. As the measured system evolves throughout a measurement interaction, then, the measuring apparatus evolves along with it; facts constitutive of one must correlate with facts constitutive of the other.

An equally significant question, though, is this: Why can't $|\psi\rangle$ evolve to become a density matrix of *nonorthogonal* vectors—that is, vectors representing nonsensical states that are not mutually exclusive and that "interfere" with each other—states whose vectors belong neither to $\mathscr{E}$ nor to $\mathscr{E}^{\perp}$ for a given fact? Such states would subsume facts that are incapable of integration, such as location in two places at once, or a cat being both alive and dead at the same time. Why doesn't one ever see such superpositions? An ontological interpretation of quantum mechanics must, it would seem, allow for their existence, since it is the evolution of these superpositions of ontologically significant potentia that quantum mechanics describes. If these potentia don't "really exist," what, exactly, is evolving mechanically? An instrumentalist would likewise answer that the indefiniteness of the initial state $|\psi\rangle$ is merely epistemological; for even if $|\psi\rangle$ could in theory evolve to become a sum of nonorthogonal vectors representing mutually interfering nonsensical states, the question as to whether or not such states can ever exist is seldom entertained in the lab, since they have never been encountered.

If, however, one seeks to explore the ontological significance of the conceptual innovations and implications of quantum mechanics—and there is clearly no reason not to do so, given that quantum mechanics has proven to be the most accurate and thus far the most fundamental description of nature yet conceived—one must be able to adequately answer these questions, and others like them. To that end, let us revisit the notion that quantum mechanics evinces two fundamental species of reality—actualities and potentia. Recall that for Heisenberg, the concept of potentia serves as "a first definition concerning the ontology of quantum theory. . . . One sees at once that this use of the word 'state,' especially the term 'coexistent state,' is so different from the usual materialistic ontology that one may doubt whether one is using a convenient terminology. . . . One may even simply replace the term 'state' by the term 'potentiality'—then the concept of 'coexistent potentialities' is quite plausible, since one potentiality may involve or overlap other potentialities."[1]

In our red-green traffic-signal example, the pure state represented by $|\psi\rangle$ can be expressed as the sum of a practically infinite number of vectors. Some of these vectors are mutually orthogonal, representing mutually exclusive potential states, and some are nonorthogonal, representing interfering or nonsensical potential states. All of these potential states are, in a sense, englobed by $|\psi\rangle$, such that $|\psi\rangle$ can be thought of as representing a sort of infusion of pure potentiality into the antecedently "actual" state of the system as it evolves. For it must not be overlooked that our traffic-signal system exists in some "actual" state prior to measurement—namely, $|u_{green}\rangle$ or $|u_{red}\rangle$—just as it will again, once its evolution, through $|\psi\rangle$, is satisfied. If, for example, our traffic signal were initially green (i.e., "prepared" in a green state prior to measurement), then $|\psi\rangle$ represents the actual state $|u_{green}\rangle$ now augmented with a pure potentiality, represented in the formalism by the potential projection of $|\psi\rangle$ upon a practically infinite number of vectors in a practically infinite number of dimensions, representing a practically infinite number of system states— some sensible, others nonsensical and seemingly impossible.

With respect to such states, Murray Gell-Mann once wrote of being assigned, as an undergraduate, the problem of using quantum mechanics to calculate the probability that a heavy macroscopic object would, during a given time interval, jump a foot into the air—a ridiculous state nevertheless subsumed by $|\psi\rangle$. "The answer," Gell-Mann wrote, "was around one divided by the number written as one followed by sixty-two zeros"[2]—an exceedingly improbable state, for all practical purposes impossible, yet far from ontologically insignificant; for the potentiality of similar, exceedingly improbable states is precisely why useful effects such as quantum tunneling are possible, and why devices that make use of these effects, such as the scanning tunneling microscope (STM), are able to function.

Returning to our idealized traffic-light system, then, it must be stressed that both $|u_{green}\rangle$ and $|\psi\rangle$ refer to the same "real" system, with the latter, potentiality-infused state $|\psi\rangle$ evolving from the former "actual" state $|u_{green}\rangle$. The systems represented by $|u_{green}\rangle$ and $|\psi\rangle$ and their evolution, one to the other, are, in other words, ontologically significant by this interpretation. According to quantum mechanics, it is the nature of such system states to evolve, and the concept of evolution, like the concept of probability, always presupposes an actual beginning and an actual end—a system in its initial state and a system in its final state—the probability of $S_{initial}$ becoming S_{final}. But evolution and probability also presuppose actuality and potentiality; both states, initial and final, are actualities—collections of facts, the former giving rise to the latter via an integration of potentia.

The primary phase of the quantum mechanical evolution of a state vector, then, is the evolution of some actual state—either $|u_{green}\rangle$ or $|u_{red}\rangle$ in our traffic light example—to the pure state $|\psi\rangle$, which like $|u_{green}\rangle$ or $|u_{red}\rangle$ is a "real" state of the system; but unlike $|u_{green}\rangle$ or $|u_{red}\rangle$, $|\psi\rangle$ can be projected onto a practically infinite number of vectors, both mutually orthogonal and nonorthogonal, thus englobing *every potential state*—both sensible and nonsensical—into which our traffic light might evolve.

At this point in the discussion, it will be useful to alter our traffic-signal example slightly so that it better resembles a typical quantum mechanical measurement interaction. To the traffic-signal system S, we shall add a detector system D, which is the instrument by which we are able to measure the color of the traffic signal. S and D together constitute a composite system whose combined Hilbert spaces form a tensor product space such that S and D together will evolve in a correlated manner, no differently than S alone as described thus far; in fact, S and D can just as well be considered individual subsystems or even individual facts of a larger system U (the universe). The dividing line separating the measured system from the measuring apparatus and environment is, as discussed earlier, purely arbitrary and a matter of convenience.

Let us suppose that our detector D is a simple pointer instrument, whose needle will point up in a state $|d_\uparrow\rangle$ if the traffic signal is red, and down in a state $|d_\downarrow\rangle$ if the traffic signal is green. Thus, the Hilbert space $\mathscr{H}_D$ representing our detector is, like $\mathscr{H}_S$, two-dimensional, and both spaces together form a single, tensor product space in which the evolution of our initial composite system $|\phi^i\rangle$ will be described.

Recall that without the detector, the pure state of S was described as:

$$|\psi\rangle = \alpha|u_{green}\rangle + \beta|u_{red}\rangle$$

Now, we have a composite, initial S and D system $|\phi^i\rangle$ whose pure state similarly evolves as:

$$|\phi^i\rangle = (\alpha|u_{green}\rangle + \beta|u_{red}\rangle)|d_\downarrow\rangle$$
$$\Rightarrow \alpha|u_{green}\rangle|d_\downarrow\rangle + \beta|u_{red}\rangle|d_\uparrow\rangle = |\phi^c\rangle$$

This evolution from the initial composite state $|\phi^i\rangle$ to the pure, correlated composite state $|\phi^c\rangle$ is described by the Schrödinger equation and constitutes the essence of a quantum mechanics; and from a practical standpoint, the correlated composite state $|\phi^c\rangle$ is entirely

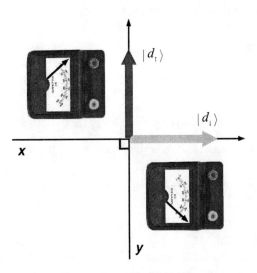

FIGURE 3.6. The two potential states D of a traffic-signal detector, represented by orthogonal vectors of unit length in a Hilbert space of two dimensions. The vector $|d_\downarrow\rangle$ along the x-axis represents "D = green" and the vector $|d_\uparrow\rangle$ along the y-axis represents "D = red."

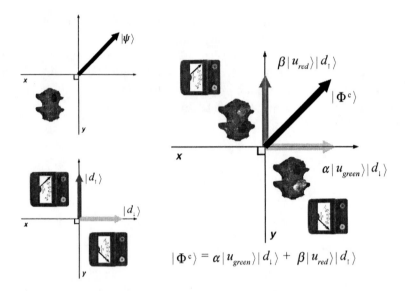

FIGURE 3.7. The composite, correlated system SD represented by the vector $|\Phi^c\rangle$.

satisfactory, since the two mutually exclusive and exhaustive, probability-valuated alternative states it subsumes, $\alpha|u_{green}\rangle|d_\downarrow\rangle$ and $\beta|u_{red}\rangle|d_\uparrow\rangle$, are both sensible: Whenever the signal is green, our detector points down, and whenever the signal is red, our detector points up. A probability is assigned to each of these alternative states, reflected in the density matrix:

$$\rho^r = |\alpha|^2|u_{green}\rangle\langle u_{green}||d_\downarrow\rangle\langle d_\downarrow| + |\beta|^2|u_{red}\rangle\langle u_{red}||d_\uparrow\rangle\langle d_\uparrow|$$

and repeated measurements will always agree with those probability valuations.

But from an ontological standpoint, we have the same difficulties discussed earlier:

(i) Why does the density matrix of the mixed state always subsume sensible, mutually exclusive and exhaustive alternative states? Why can't the mixed state ever subsume nonsensical states such as $\alpha|u_{green}\rangle|d_\uparrow\rangle$ or superpositions of states?

(ii) Superpositions aside, why does the pure state always evolve to the preferred basis of the *detector* in the mixed state, rather than some other potential state?

With respect to the latter question, it was discussed earlier that quantum mechanics does not merely describe the mutual interrelations among all facts comprised by a given system or composite system; quantum mechanics describes, rather, the evolution of the state of a system *relative to a particular fact (or subsystem) belonging to the composite system*—or, more accurately, relative to the interrelations between this particular fact or subsystem and all other facts belonging to the measured system. In quantum mechanics, this "particular fact or subsystem"—referred to as the "indexical eventuality"— always belongs to the measuring apparatus. Thus, the mixed state always reflects the preferred basis of the measuring apparatus, since the state of the system evolves relative to the indexical eventuality belonging to this apparatus. The grouping of potential state vectors into subspaces $\mathscr{E}$ and $\mathscr{E}^{\perp}$ exemplifies this relativity, where all potential states sharing the indexical eventuality $|d_\uparrow\rangle$, for example, belong to $\mathscr{E}$ and all others belong to $\mathscr{E}^{\perp}$. It is this evolution of the pure state relative to a particular fact (or subsystem) that conditions the elimination of superpositions of interfering, "coherent" states via a process of negative selection. The only potential states that remain—those constituting the mixed state—are, as a result, noninterfering, "decoherent" potential states, which have been valuated as probabilities.

Before exploring the mechanics of this elimination via negative selection, effected by the state's evolution relative to a particular fact, it should be pointed out that this mechanism is often described more simply as follows: "Superpositions are eliminated by the measuring apparatus as it interacts with the system measured." Heisenberg, for example, writes that in the quantum formalism, the superposition or interference of potentia, "which is the most characteristic phenomenon of quantum theory, is destroyed by the partly indefinable and irreversible interactions of the system with the measuring apparatus and the rest of the world."[3] The difficulty of this wording, with its arbitrary separation of the universe into system, apparatus, and "the rest of the world" (environment), is the intrusion of sheer subjectivity it implies; for as we saw earlier, the final "measuring apparatus" in the von Neumann chain of apparatuses is, unavoidably, the mind of the human observer. In defense against this implication, Heisenberg states:

Certainly quantum theory does not contain genuine subjective features, it does not introduce the mind of the physicist as part of the atomic event. But it starts from the division of the world into the "object" and the rest of the world. . . . This division is arbitrary and historically a direct consequence of our scientific method; the use of classical concepts is finally a consequence of the general human way of thinking.[4]

Since, however, we are here exploring an ontological rapprochement of the classical, materialistic worldview and a worldview capable of accommodating quantum mechanics, it is important to acknowledge explicitly the following two points: (i) the universe, since it is the only truly closed system, is the only system to which quantum mechanics can properly apply *if* quantum mechanics is to be thought of as being ontologically significant; and (ii) the grouping of various facts within this system into composite subsystems of "system," "apparatus," and "environment" subsystems is purely arbitrary. Given these two points, one should likewise acknowledge that the evolution of the state of a measured system is, in fact, the evolution of the state of the universe itself; therefore, its evolution relative to a given "measuring apparatus" is merely its evolution relative to a particular fact (or collection of facts) belonging to itself. In the same sense, the distinction between a single fact and a subsystem of facts within the universe is arbitrary as well, and this applies to the states S (system), D (detector), and E (environment). The correlation among these subsystems is easily comprehended by virtue of the fact that as the closed system of the universe evolves, so must all the facts subsumed by it, however they might be grouped and whatever they might be named. These last two points are clearly expressed by Heisenberg in the following:

The measuring device . . . contains all the uncertainties concerning the microscopic structure of the device which we know from thermodynamics, and since the device is connected with the rest of the world, it contains in fact the uncertainties of the microscopic structure of the whole world.[5]

These ontological implications aside, however, the quantum mechanical description of a measurement interaction, such as the one in our example, is typically left at:

$$|\Phi^i\rangle = (\alpha|u_{green}\rangle + \beta|u_{red}\rangle)|d_\downarrow\rangle$$
$$\Rightarrow \alpha|u_{green}\rangle|d_\downarrow\rangle + \beta|u_{red}\rangle|d_\uparrow\rangle = |\Phi^c\rangle$$

since this expression provides sensible outcome states, each of which is accurately valuated as a probability, such that successful predictions can be made. The question as to the mechanism of negative selection by which the superpositions of interfering states in $|\psi\rangle$ are eliminated from the above expression is a question which begs the ontological significance of quantum mechanics, and thus has received relatively little attention in physics.

DECOHERENCE

The persisting lack of a truly coherent ontology capable of accommodating quantum mechanics, however, has throughout recent years inspired a number of physicists to examine more closely this mechanism of negative selection by which quantum superpositions are eliminated. It was stated earlier that the key to this mechanism, for certain physicists, was to be found in the very desiderata required by a coherent ontological interpretation of quantum mechanics, with the most significant of these being that the interpretation be universally applicable. The closed system in which every quantum mechanical measurement takes place must, at a fundamental level, be understood to be the universe itself; and from that stipulation, it must be acknowledged that the environment englobing a measured system (i.e., a measured subsystem of the universe)—though it may be and typically must be ignored for the practical purposes of calculation—must therefore play some role in the evolution of the state of the measured system. Many theorists, including Wojciech Żurek, Robert Griffiths, Roland Omnès, James Hartle, and Murray Gell-Mann, among several others, have demonstrated that the interrelations between "measured" facts and facts belonging to the "environment" englobing the measurement interaction play a crucial mechanical role in the elimination of superpositions via negative selection. Though the proposals of each of these thinkers differ somewhat, they all emphasize the importance of this negative selection process, and mutually refer to the result of this process as the "decoherence effect."

In other words, the correlated superposition

$$|\Phi^c\rangle = \alpha|u_{green}\rangle|d_\downarrow\rangle + \beta|u_{red}\rangle|d_\uparrow\rangle$$

evolves via a process of negative selection to become a mixed state of *decoherent*, probability-valuated, mutually exclusive and exhaustive alternative states in the reduced density matrix:

$$\rho^r = |\alpha|^2 |u_{green}\rangle\langle u_{green}| |d_\downarrow\rangle\langle d_\downarrow| + |\beta|^2 |u_{red}\rangle\langle u_{red}| |d_\uparrow\rangle\langle d_\uparrow|$$

Furthermore, these theorists agree that the negative selection mechanism by which decoherence is effected involves the environment of the system measured—though they disagree slightly as to how, exactly, the environment functions in this regard. The extent of this disagreement lies beyond the scope of this essay, however, so for the current discussion it will suffice to focus on just one of these decoherence interpretations—the Environmental Superselection interpretation of Wojciech Żurek.[6]

"Decoherence results from a negative selection process that dynamically eliminates nonclassical states," writes Żurek, who maintains that decoherence is a consequence of the universe's role as the only truly closed system—which, put another way, guarantees the ineluctable "openness" of every subsystem within it. "This consequence of openness is critical in the interpretation of quantum theory," Żurek continues, "but seems to have gone unnoticed for a long time."[7] The key to understanding the negative selection mechanism of decoherence in Żurek's interpretation is the supplemental process, suggested by von Neumann and introduced in chapter 2[8]—his "Process 1"—whereby the reduced density matrix of the mixed state evolves from a larger, correlated, pure state density matrix. This larger, pure state density matrix, unlike the reduced matrix, contains nonsensical, coherent states (noted in boldface) which must be eliminated:

$$\rho^c = |\alpha|^2 |u_{green}\rangle\langle u_{green}| |d_\downarrow\rangle\langle d_\downarrow| + \boldsymbol{\alpha\beta^*} |u_{green}\rangle\langle \mathbf{u_{red}}| \mathbf{d_\downarrow}\rangle\langle \mathbf{d_\uparrow}|$$
$$+ \boldsymbol{\alpha^*\beta} |\mathbf{u_{red}}\rangle\langle u_{green}| \mathbf{d_\uparrow}\rangle\langle \mathbf{d_\downarrow}| + |\beta|^2 |u_{red}\rangle\langle u_{red}| |d_\uparrow\rangle\langle d_\uparrow|$$

This evolution of the state vector from the correlated pure state $|\phi^c\rangle$ and its pure state density matrix ρ^c to the mixed state and its reduced density matrix ρ^r is useful, in that one is able to account for the elimination from ρ^c of the nonsensical, correlated states that are incapable of integration in the reduced density matrix. The "off-diagonal" terms in boldface represent these nonsensical states, whose vectors are mutually nonorthogonal. Typically, one merely cancels out these off-diagonal terms, yielding the reduced density matrix; however, the ontological interpretations that make use of the decoherence effect suggest an ontologically significant process by which the interaction of the measured system with its measuring apparatus and its environment produces the necessary cancellations.

In Żurek's Environmental Superselection interpretation,[9] the interaction of the environment $|E_0\rangle$ with $|\phi^c\rangle$ is described as:

$$|\Phi^c\rangle|E_0\rangle = (\alpha|u_{green}\rangle|d_\downarrow\rangle + \beta|u_{red}\rangle|d_\uparrow\rangle)|E_0\rangle$$
$$\Rightarrow \alpha|u_{green}\rangle|d_\downarrow\rangle|E_\downarrow\rangle + \beta|u_{red}\rangle|d_\uparrow\rangle|E_\uparrow\rangle = |\Psi\rangle$$

This correlation of the environment, the system measured, and the detector, and the cancellations of interfering nonorthogonal potential states it produces can most easily be visualized by recalling the simple concept of subspaces: The subspace $\mathscr{E}_\downarrow$ contains all vectors representing all potential states of the universe where our detector's

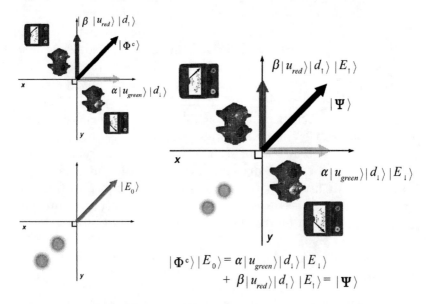

$$|\Phi^c\rangle|E_0\rangle = \alpha|u_{green}\rangle|d_\downarrow\rangle|E_\downarrow\rangle$$
$$+ \beta|u_{red}\rangle|d_\uparrow\rangle|E_\uparrow\rangle = |\Psi\rangle$$

FIGURE 3.8. The composite system SD is correlated with the Environment (E), forming a new correlated, composite system SDE, represented by the vector $|\Psi\rangle$.

pointer is down: $|d_\downarrow\rangle$. Recall that a state is a maximal specification of all facts comprised by a system, so that every potential state vector belonging to $\mathscr{E}_\downarrow$ represents a different, potential "snapshot" of the universe and all its facts. But in $\mathscr{E}_\downarrow$ all of these snapshots have at least one fact in common: Our detector's pointer is down. Likewise, we have a subspace consisting of all vectors orthogonal to $\mathscr{E}_\downarrow$ and in this subspace $\mathscr{E}_\downarrow^\perp$ the fact common to all vectors belonging to it is that our detector's pointer is not down.

Because the environment subsumes a nearly infinite number of potential facts, its correlation with our system-detector $|\phi^c\rangle$ produces a practically infinite number of degrees of freedom—different potential projections of $|\Psi\rangle$ upon $\mathscr{E}_\downarrow$ and $\mathscr{E}_\downarrow^\perp$, and $\mathscr{E}_\uparrow$ and $\mathscr{E}_\uparrow^\perp$, producing a practically infinite number of potential state vectors, some of which belong to $\mathscr{E}_\downarrow$ and some of which belong to $\mathscr{E}_\downarrow^\perp$.

This unimaginably large number of potential state vectors in $\mathscr{E}_\downarrow$ and $\mathscr{E}_\downarrow^\perp$ is useful because it produces a great deal of cancellation of interfering, incompatible states within each of these subspaces.

In an ontological interpretation of quantum mechanics for which logical consistency is a desideratum, the justification for this conception of 'interfering, incompatible states' and the elimination of such states is simply a matter of satisfying the logical principles of non-contradiction and the excluded middle. While it is true that many physicists tend to dismiss such ontological justifications as unscientific, metaphysical hand-waving, it must be remembered that the correlation of causal relation and logical implication, central to modern science, is a *presupposed* correlation; it is, in other words, an ontological 'first principle' of science. Since modern science is impossible without this presupposition of logical causality, the appeal to logical causality as an ontological desideratum of quantum mechanics would seem entirely justified as a necessary and sufficient principle to account for the elimination of superpositions of interfering states.

Consider, for example, the following two potential states of $|\Psi\rangle$ belonging to $\mathscr{E}_\downarrow$:

(i) A state where our detector points down, our traffic signal is green, and a particular photon from the sun strikes the green lens.

(ii) A state where our detector points down, our traffic signal is red, and this particular photon strikes the red lens.

These two potential states are interfering states, incapable of integration, as are the vast majority of states in $\mathscr{E}_{\downarrow}$ and $\mathscr{E}_{\downarrow}^{\perp}$, and serve to cancel each other out. In our experiment, we are concerned only with the correlation of the detector and the traffic signal, such that we ignore all other potential facts in the universe, including the status of photons from the sun as they strike the traffic signal. But because these and the multiplicity of other facts belonging to the environment are correlated with our system and detector, the cancellations they produce also eliminate the interferences we are concerned with, such as our detector pointing down and our traffic signal being red instead of green.

Put another way, the Hilbert spaces describing our traffic signal and detector by themselves span only two dimensions, which are not sufficient to produce the cancellations needed to eliminate unwanted superpositions of interfering states. By correlating our system and detector with the environment and its Hilbert space of practically infinite dimensions, we are able to produce enough potential states of $|\Psi\rangle$ such that most of these are mutually interfering and self-cancelling, thereby eliminating superpositions of interfering states.

This cancellation is represented mathematically as a "trace-over" of the uncontrolled and unmeasured degrees of freedom contributed by the environment, and the density matrix that results from this cancellation is written as:

$$\rho_{SD} \equiv \mathrm{Tr}_E |\Psi\rangle\langle\Psi| = \Sigma_i \langle E_i|\Psi\rangle\langle\Psi|E_i\rangle = \rho^r$$

with ρ^r here being identical to the reduced density matrix we had earlier, before introducing the environmental correlations:[10]

$$\rho^r = |\alpha|^2 |u_{green}\rangle\langle u_{green}| |d_{\downarrow}\rangle\langle d_{\downarrow}| + |\beta|^2 |u_{red}\rangle\langle u_{red}| |d_{\uparrow}\rangle\langle d_{\uparrow}|$$

Thus, the interrelation of environmental facts with the facts comprised by our system and detector continuously reduces any coherent, interfering superpositions of states relative to the preferred basis $|d_{\downarrow}\rangle$ and $\langle d_{\uparrow}|$, such that the only alternative states in the reduced density matrix are mutually exclusive and exhaustive, logical states, each valuated according to a probability. "An effective superselection rule has emerged," Zurek writes. "Decoherence prevents superpositions of the preferred basis states from persisting. Moreover, we have obtained all this—or so it appears—without having to appeal to anything beyond the ordinary, unitary Schrödinger evolution."[11] (See Figure 3.9 on front end sheet.)

The ontological significance of this interpretation of quantum mechanics, with its mechanism of negative selection of interfering potentia via the decoherence effect, becomes acute when one explores the cosmological implications—especially since this interpretation of quantum mechanics explicitly recognizes that all facts subsumed by the closed system of the universe are mutually interrelated in any given measurement interaction. Many theorists have pursued these implications, and in so doing have suggested a more intuitive and more generalized way of describing the evolution of states of the universe: Instead of focusing on states (akin to "snapshots" of the actualities of the universe), these theorists focus on "histories," or historical routes of states—that is, histories of facts whose perpetual augmentation by novel facts is presupposed.

The elimination of interfering potentia and the probability-valuation of those that remain, via the environmental correlations and trace-over described above, is thus seen as the elimination of *interfering potential histories*. Separate systems with separate histories, by virtue of their correlation with a shared environment (i.e., the universe), necessarily entail many shared potential facts. And since these shared potential facts must be mutually consistent, so must those histories which include them—lest nature become overrun with violations of the logical principle of noncontradiction (physical bodies being two places at once, etc.)—the fundamental principle which bridges the logical realm of the necessary and the causal realm of the contingent. Interfering potential histories, those that are not mutually exclusive and exhaustive, are thus eliminated by integration via environmental correlation. Thus, the mutual historical consistency of multiple systems, even those which are spatially well separated, derives in large part from their being englobed by a larger, shared "environmental history."

This concept of decoherent, consistent histories associated with environmental correlations is central to the quantum mechanical interpretations of Robert Griffiths,[12] Roland Omnès,[13] and Murray Gell-Mann and James Hartle.[14] In all of these interpretations, the term "history" implies an ongoing process not necessarily connoted in the term "state." For in quantum mechanics, as we have seen, it is the nature of a state to evolve continuously in one sense, yet discontinuously in another, just as a history of actual events evolves;

thus, the evolution of both "state" and "history" is characterized as fluid potentiality punctuated by an ever-expanding lineage of discrete actualities/events/facts, each novel actuality being subsequent to and partially consequent of all actualities antecedent to it. And just as the concepts of probability and the evolution of a state both imply an actual past and a presupposed, subsequent, and consequent actual future, so does the concept of a "history."

Thus, we may characterize the alternative states of the reduced density matrix of any given measurement interaction as "alternative potential histories" of the universe. This set of histories is understood to be interrelated with all antecedent facts belonging to the universe, and it is this interrelation which allows for a subsequent elimination of the vast majority of potentia produced by this interrelation, such that interfering superpositions of potential histories—histories incompatible for integration in the reduced density matrix—are eliminated. Those histories that remain are referred to as "consistent histories"—the definition of which, as applied to quantum mechanics, was first proposed by Robert Griffiths,[15] who demonstrated that not only integrations of alternative potential states but also *sequences* or histories of alternative potential states can eliminate interfering states. Griffiths thus demonstrated the equivalence between alternative states and alternative histories of states valuated as mutually exclusive and exhaustive probabilities in the reduced density matrix. Gell-Mann and Hartle have shown that Griffiths's "consistency conditions" are similar to those required for the decoherence effect, with the conditions for the latter being slightly stronger, in fact (hence, their use of the term "strong decoherence," discussed in their paper of the same name[16]).

THREE BENEFITS OF THIS ONTOLOGICAL INTERPRETATION OF THE QUANTUM FORMALISM

Though there are a great many benefits to the ontological interpretation of quantum mechanics presented thus far, three stand out at this point in the discussion. First, this interpretation is a truly universal interpretation of quantum mechanics. Unlike models based upon Bohr's notion of complementarity, with its arbitrary divisions of the world into "subject" and "object," each with its own body of physical laws incompatible with those of the other, the decoherence-based interpretations require no such division; rather, these interpretations make use of the principle that facts belonging to "subject," "object," and "environment" are necessarily ontologically interrelated. These interrelations give rise, via the Schrödinger equation, to manifold logical integrations of potential states which, via nothing other than the process of integration itself, eliminate superpositions of interfering states incompatible for integration. The decoherence-based interpretations thereby purge quantum mechanics of its infamous predication of objective reality upon subjective—and by some interpretations, necessarily human—observations. It is because of this first benefit that decoherence-based interpretations of quantum mechanics are often described as "quantum mechanics without observers." Of course, this characterization is not entirely accurate insofar as "observation" by this interpretation is described as the interrelation among all facts subsumed by a system—that is, observer-facts interrelated with measured system-facts and environmental facts. The advantage of the decoherence-based interpretations is that they explicitly acknowledge and make use of the concept that the universe itself is the only truly closed system, and therefore *all* facts englobed by this system are necessarily mutually interrelated and ontologically incapable of isolation; and by this principle, the role of the "observer" in orthodox quantum mechanics is always fulfilled.

In fact, it has been suggested by Gell-Mann and Hartle that the mostly classical or "quasi-classical" behavior of the relatively high-inertia macroscopic objects we typically encounter in everyday experience is, in part, a product of the decoherence effect; but it is also a product of the extremely high degree of historical correlation among facts—that is, reproduction from fact to fact along an historical route—typical of high-inertia systems, such that it is always potentia with probability valuations extremely close to one that are reproduced with great regularity along a given historical route of facts. This tendency toward potentia regulated by reproduction is typical of the high-inertia systems that constitute most of our everyday world. Schrödinger's cat, for example, is a sufficiently large enough system of facts that its own "internal environment"—apart from the unavoidable coupling with the environment "external" to the cat (the air it breathes interacting with the box it is in, interacting with the world outside the box, etc.)—easily provides the requisite degrees of freedom to produce the necessary cancellations of interfering potential states. Those potential states that survive the cancellations are highly regulated by closely correlated reproductions among facts constituting the historical route of the cat. The result is that if the cat is dead at time t, there is a probability extremely close to one that it will be dead at time $t + 1$ as well. These two factors combined—decoherence, and highly regulated reproduction among facts comprising high-inertia systems—offer a clear ontological account as to why we never see live-dead cat superpositions, or any other macroscopic superpositions even though such superpositions are a necessary component of quantum mechanics. Bluntly stated by Gell-Mann: "Reams of paper have been wasted on the supposedly weird quantum-mechanical state of the cat, both dead and alive at the same time. No real quasiclassical object can exhibit such behavior because interaction with the rest of the universe will lead to decoherence of the alternatives."[17]

The second benefit of this conception of system states as histories of actualities/events/facts is that it allows one to intuitively see how the so-called paradox of quantum nonlocality—arguably the most formidable obstacle to a strictly classical accommodation of quantum mechanics—is really no paradox at all. Quantum nonlocality is simply the correlation among facts in a composite quantum system, such as our example system

$$|\Psi\rangle = \alpha|u_{green}\rangle|d_\downarrow\rangle|E_\downarrow\rangle + \beta|u_{red}\rangle|d_\uparrow\rangle|E_\uparrow\rangle$$

when S, D, and E entail actualities that are spatially well separated. For according to the formalism above, facts belonging to E, even though they may pertain to a part of the universe far removed from S and D, nevertheless affect the actualization of potential facts belonging to S and D.

The problem of quantum nonlocality was first formally addressed in the famous paper of Einstein, Podolsky, and Rosen (EPR) in 1935, and in 1969 an experiment was designed by Abner Shimony, John Clauser, M. Horne, and R. Holt to test the EPR argument and its newer formulation by David Bohm—that supposedly nonlocal correlations are merely statistical artifacts that derive from hidden variables, and that if these variables were disclosed, quantum mechanics would be a complete, fully deterministic theory. In 1964, John Bell demonstrated mathematically that the nonlocal correlations predicted by quantum mechanics exceeded those predicted by classical, statistical mechanics;[18] his theorem was tested experimentally in 1972 by Clauser and S. Freedman,[19] and in 1982 another version of the experiment was performed by A. Aspect, J. Dalibard, and G. Roger.[20] The results of these experiments vindicated quantum mechanics by their demonstration of nonlocal interrelations in excess of those allowed for by classical statistical mechanics—to the surprise of many. In the years since these experiments, a great deal has been written about this supposedly bizarre, counterintuitive, and conceptually problematic nonlocality feature of quantum mechanics.

The interpretive strategy represented by the arguments of Einstein, Podolsky, Rosen, and Bohm entails the characterization of quantum nonlocality—as well as indefiniteness, chance, entanglement, probability—as epistemic artifacts of the quantum theory's inability to account for a multiplicity of "hidden variables" that, if disclosed, would complete the theory. Thus, quantum mechanics would be cleansed of all uncertainty so that its true essence as a fully determinate, classical theory would be revealed. Since human limitations prevent the specification of these deterministic hidden variables, however, the incompleteness of the theory with its unmovable veil of indeterminism relegates us to a statistical approximation of them. Thus the matrix of probable outcome states yielded by quantum mechanics is to be interpreted, as advocated early on by Max Born, as statistical probabilities of purely epistemic significance[21]—probabilities that describe the statistical frequency at which an experimenter measuring the position of an electron, for example, finds it in a given region around an atomic nucleus after repeated experiments. It is, in other words, the probability that one will find that the *pre-established facts* will fit a *probable form* a certain percentage of the time when a given experiment is repeated sufficiently.

This statistical interpretation of quantum mechanics is thus able to account for indefiniteness, change, probability, and entanglement by rendering these concepts merely epistemically significant, as opposed to ontologically significant. But the statistical or "hidden variables" interpretation of quantum mechanics is not so easily able to account for nonlocality; for even if nonlocal correlations between spatially well separated systems are interpreted as statistical artifacts owing to hidden variables—which, for example, might pertain to the stochastic fluctuations of the field of atomic and subatomic particles linking the two systems—the classical interactions among these particles are still limited by the speed-of-light boundary of special relativity. Any causally significant transference of energy from one system to the other along the field—whether statistically veiled by its (apparently) stochastic fluctuations, or whether explicitly apparent as in a communication signal beamed from one system to the other—cannot exceed the speed of light. But the experiments of Aspect et al. reveal that if nonlocality is to be explained according to this classical deterministic model, these propagations along the field must be superluminal.

In answer to this difficulty, many advocates of the experimentally disconfirmed classical, "local hidden variables" interpretation moved to embrace a "nonlocal hidden variables" interpretation, which is otherwise the same in its classical interpretation of quantum mechanics—except that it allows for nonlocal interactions via a superluminal propagation of energy, in violation of special relativity. The nonlocal hidden variables theory of David Bohm, introduced in 1952,[22] is the most complete and systematized incarnation of this type of theory; in it, Bohm suggests that the entire universe is permeated by an ether of point particles constitutive of an "implicate order" that can be described and accounted for via quantum mechanics. This not only allows for nonlocal interactions via superluminally propagated energy, characterized as a kind of "shock wave," similar to the "pilot wave" model proposed by de Broglie in 1927; it also renders the universe fundamentally time-reversal invariant—that is, fundamentally deterministic, such that past, present, and future are all mutually internally related.

To illustrate this fundamental deterministic symmetry of Bohm's "implicate order," I refer again to the experimental device described by Bohm and his colleague B. J. Hiley referenced in chapter 1:

> This device consists of two concentric glass cylinders; the outer cylinder is fixed, while the inner one is made to rotate slowly about its axis. In between the cylinders there is a viscous fluid, such as glycerine, and into this fluid is inserted a droplet of insoluble ink. Let us now consider what happens to a small element of fluid as its inner radius moves faster than its outer radius. This element is slowly drawn out into a finer and finer thread. If there is ink in this element it will move with the fluid and will be drawn out together with it. What actually happens is that eventually the thread becomes so fine that the ink becomes invisible. However, if the inner cylinder is turned in the reverse direction, the parts of this thread will retrace their steps. (Because the viscosity is so high, diffusion can be neglected.) Eventually

the whole thread comes together to reform the ink droplet and the latter suddenly emerges into view. If we continue to turn the cylinder in the same direction, it will be drawn out and become invisible once again.

When the ink droplet is drawn out, one is able to see no visible order in the fluid. Yet evidently there must be some order there since an arbitrary distribution of ink particles would not come back to a droplet. One can say that in some sense the ink droplet has been enfolded into the glycerine, from which it unfolds when the movement of the cylinder is reversed.

Of course if one were to analyse the movements of the ink particles in full detail, one would always see them following trajectories and therefore one could say that fundamentally the movement is described in an explicate order. Nevertheless within the context under discussion in which our perception does not follow the particles, we may say this device gives us an illustrative example of the implicate order. And from this we may be able to obtain some insight into how this order could be defined and developed.[23]

As regards the phenomenon of quantum nonlocality, Bohm and Hiley suggest a modification of this model:

> Suppose that we consider two particles initially at different positions, one represented by red ink and the other by blue ink. As we enfold these into the fluid, red ink particles and blue ink particles will eventually intermingle, so that in a given region we could not say that the ink droplets were really separate. Nevertheless when the cylinder was turned back, red and blue ink particles would begin to separate to form respectively red and blue droplets. It is clear that the ground of the particular behaviour of the visible droplets is global and that disturbances in this ground can lead to correlated changes in this behaviour.
>
> In our model the appearance in perception of what seems to be a pair of separate ink droplets can therefore be deceiving. Actually these apparently separate ink droplets are internally related to each other. On the other hand, as conceived in the explicate order, such droplets would only be externally related. This is one of the most important new features of the implicate order. What may be suggested is that this feature may give further insight into the meaning of quantum properties such as mutual participation and nonlocality.[24]

Again, the mechanism by which the implicate order is mediated between the nonlocal actualities is a "pilot wave" propagated throughout the ether of hidden point particles constituting the implicate order. But this method of mediation is not compatible with special relativity because it is not Lorentz invariant. Bohm's interpretation requires, in other words, that spatially well separated systems be connected instantaneously (and therefore superluminally) by the pilot wave. The deterministic implications of this requirement were explored in chapter 1, as was the related implication of fundamental temporal symmetry. With respect to the concept of instantaneous nonlocal interactions, however, these implications become intimately bound up in the possibility of faster-than-light communication. Given the pilot wave's instantaneous connection of spatially well separated systems in Bohm's theory, in violation of special relativity, superluminal communication should be possible; that it has not yet been achieved is, according to Bohm, primarily a consequence of experimental limitations. He writes:

> Of course, until we are able to carry out experiments on individual processes that are more accurate than the limits set by the uncertainty principle, it will not be possible to use quantum nonlocality for the purpose of sending signals. . . . Any attempt to send a signal by influencing one of a pair of particles under EPR correlation will encounter the difficulties arising from the irreducibly participatory nature of all quantum processes. If for example we tried to 'modulate' the overall wave function so that it could carry a signal in a way similar to what is done by a radio wave, we would find that the whole pattern of this wave would be so fragile that its order could change radically in a chaotic and complex way. As a result no signal could be carried.[25]

The ether of point particles, in other words, is stable enough to produce EPR-like nonlocal correlations superluminally, but not stable enough to produce a communicative signal superluminally. This difficulty aside, the central point here is that in Bohm's hidden variables theory, the experimentally verifiable phenomenon of quantum nonlocality is accounted for by a dynamical mechanism which is fundamentally incompatible with special relativity. When one recalls that the original point of the hidden variables interpretations of quantum mechanics was to render it a classically deterministic theory, the instantaneous nonlocal connection of particles by a Lorentz noninvariant, superluminal pilot wave seems a serious price to pay—the salvaging of fundamental classicality by means of selective dispensations from special relativity, one of its most central principles.

By contrast, quantum nonlocality, when interpreted via a satisfactory ontological framework such as that proposed by Żurek, Omnès,

Gell-Mann, et al., is entirely reasonable and classically intuitive—so much so, in fact, that one can use "real world" situations to demonstrate this interpretation of quantum nonlocality (a good indicator of the coherence of this ontological framework). The following quantum nonlocality analogy is an example of what is sometimes referred to as a "Cambridge change"—a term that likely derives from P. T. Geach's use of it in his 1969 book *God and the Soul*.[26] The term refers to Cambridge philosophers such as Bertrand Russell and J. M. E. McTaggart, who had considered the idea of changing descriptive qualifications (e.g., per this discussion, "changing quantum mechanical states") as ontologically significant to the object of the description. Geach writes:

> The only sharp criterion for a thing's having changed is what we may call the Cambridge criterion (since it keeps on occurring in Cambridge philosophers of the great days, like Russell and McTaggart): The thing called "*x*" has changed if we have "F(*x*) at time *t*" true and "F(*x*) at time *t*¹" false, for some interpretation of "F", "*t*", and "*t*¹". But this account is intuitively quite unsatisfactory. By this account, Socrates would after all change by coming to be shorter than Theaetetus; moreover, Socrates would change posthumously (even if he had no immortal soul) every time a fresh schoolboy came to admire him; and numbers would undergo change whenever e.g. five ceased to be the number of somebody's children.
>
> The changes I have mentioned, we wish to protest, are not "real" changes; and Socrates, if he has perished, and numbers in any case, cannot undergo "real" changes. I cannot dismiss from my mind the feeling that there is a difference here; and I suggest that when we have a narrative proposition corresponding to a "real" change, there is individual actuality—an imperfect actuality, Aristotle calls it—that is the change; but not, when a mere "Cambridge" change is reported. (Of course there is a "Cambridge" change whenever there is a "real" change; but the converse is not true.) But it would be quite another thing to offer a criterion for selecting, from among propositions that report at least "Cambridge" change, those that also report "real" change (given they are true); and I have no idea how I could do that.[27]

The decoherence or "consistent histories" interpretations of quantum mechanics provide an excellent solution to the question of such Cambridge changes, even those that entail nonlocal correlations. The key, as we have seen, lies in two fundamental principles: First, the concept of ontologically significant potentia, and the understanding that actualities (what Geach refers to as "real" changes) and potentia constitute the two fundamental species of reality; and second, the fundamental characterization of objects as serial, historical routes of quantum events—quantum actualizations. Thus, as the *potentia* associated with an historical route of quantum mechanical state specifications change, then in some sense the system described by this historical route must also change. The potentia associated with a history, in other words, contribute to its definition in an ontologically significant way, just as the actualities associated with a history do; the difference is that the potentia associated with a quantum mechanical history can be affected nonlocally by correlation with the actualities of other histories. All that is required is the inclusion of the correlated histories in some broader, shared environmental history by which logically necessary consistency conditions become operative. These consistency conditions are ultimately grounded in the logical principles of noncontradiction and the excluded middle, the foundation of any and all logical constructions, including, obviously, the mathematics that lie at the heart of quantum theory.

Consider, for example, the history of actualities/events/facts comprised by a traveling salesman, whose pregnant wife in California is soon to deliver. Certain of the actualities/events/facts belonging to the salesman pertain to his location, which is somewhere in Hong Kong, far away from his wife. Nevertheless, the history describing the salesman is correlated with the history describing his wife, such that as soon as she gives birth, we can say that the salesman's history has been augmented in a definite, objective way, evinced by asking the simple question: When does the salesman become a father? When, in other words, are the potentia *objectively pertinent* to his future affected? According to a classical mechanical interpretation of this situation, the answer would be: As soon as his wife is able to communicate this fact to him—communication limited to the speed of light, according to special relativity.

But according to a quantum mechanical interpretation of this situation, the history of states describing the salesman instantly changes as soon as his wife gives birth, by virtue of the fact that his state history is correlated with that of his wife. The correlations of their two histories are accounted for by two concepts: (i) Both histories are correlated with an environmental history, which is itself englobed by the universal history $|\Psi\rangle$, and therefore all these correlated histories must be mutually consistent; (ii) the consistency among the salesman and wife histories is a function of the decoherence effect, which eliminates any incompatible, interfering histories in the pure state density matrix, such that the alternative histories comprising the reduced state density matrix are consistent, mutually exclusive and exhaustive probabilities.

The fact that the history of states describing the salesman changes as soon as his wife gives birth does not, as has been suggested in some discussions of the real experiments involving quantum nonlocality, imply any "superluminal communication" in violation of special relativity. Theoretical exploration of the possibility of such faster-than-light communication via quantum nonlocality has revealed that correlations among nonlocal histories cannot involve the communication of information. Put another way, quantum mechanics allows the actualities of one system to causally *affect* the *potentia* of a spatially well separated system by virtue of the correlated histories of the two systems; quantum mechanics does *not*, however, allow spatially well separated systems to *influence* one another causally. This distinction between "causal affection of potentia by logically prior actuality" and "causal influence of actualization by temporally prior actuality" is perhaps negligible in classical mechanics, where both constitute the qualification of material substance by quality. In an ontological interpretation of quantum mechanics, however, the distinction between causal affection of potentia and causal influence of actualization is directly related to and necessitated by the ontological significance of potentiality. For as the evolution of a state or history of a system is driven by the potentia associated with the system—potentia that are subsequently and consequently related to all logically prior facts in the closed system of the universe—an *affection* upon these potentia does not necessarily entail causal *influence* upon the quantum mechanical actualization of the system.

The state history of the salesman, in other words, involves not only the actualities/events/facts subsumed by his state history as it *exists*, but also the *potential* actualities/events/facts subsumed by his state history as it *evolves*; and it is these potentia that are affected nonlocally. Though he is not aware of his state change, for the same reason superluminal communication is not possible in quantum mechanics, the potentia associated with the salesman's *subsequent* and *consequent* history—and the role of these potentia in forming the probabilities for their actualization—are affected, and in an ontologically significant way. One sees here the rehabilitation of the Aristotelian intuition that a thing is not only defined by what it was in the past, but also by what it might become in the future—an intuition we typically restrict to those occasions wherein we define ourselves. The decoherent histories interpretations of quantum mechanics, however—interpretations which stress that quantum mechanics predicts *probable potential* outcome states and not *determined actual* outcome states—reveal that it is an intuition which we might sensibly apply to the whole of nature.

The intuitive ontological concepts discussed in these real-world examples of nonlocality are identical to those used in interpreting quantum nonlocality in the famous EPR-type laboratory experiments performed by Aspect et al., which is but one indicator of the coherence of the ontological interpretation of quantum mechanics presented thus far.

The third benefit of this interpretation, closely related to nonlocality, is that it mediates the controversy in physics, originating in Boltzmann's kinetic theory, between (i) the temporal symmetry implied by the time-reversal invariance of classical dynamics and (ii) the temporal asymmetry—the one-way directionality of time—exemplified by the second law of thermodynamics. Since it is fundamentally the time-independent form of the Schrödinger equation that describes the evolution of the state of a system (or a history of such states) from an actual state to an integration of potential states, this evolution essentially occurs "out of time." It is the actualization of one of these potential states—the discontinuous punctuation of this symmetrically atemporal continuum of potentia—which results in an asymmetrical temporality—the only temporality we know. By this ontological interpretation of quantum mechanics, then, asymmetrical time is in fact a *product* of the actualizations of potentia, rather than a background against which such actualizations occur.[28] One manifestation of this is the question: "What happens to an electron in between observations?" Heisenberg answers that quantum mechanics "does not allow a description of what happens between two observations. Any attempt to find such a description would lead to contradictions; this must mean that the term 'happens' is restricted to observation."[29] Asking, "What happens to an electron in between observations?" is, by the decoherence-based interpretations of quantum mechanics, the same as asking, "What happens to an electron in between actualizations?" The answer, of course, is that nothing "happens" to it (apart from its own internal quantum mechanical state evolution) since the electron does not exist as an actual entity in between actualizations; it exists only as a potential entity in between actualizations. The only electrons that are "actual" are those of the past relative to an evolving actualization in the present. The past is thus settled and the future remains open, with subsequent actualizations causally influenced by the objective data of past actualities, but not determined by these data.

Temporal asymmetry is crucial to an ontological interpretation of quantum mechanics, since logically ordered causal dynamics are impossible apart from such asymmetry; but temporal asymmetry is neither provided nor accounted for by the Schrödinger equation in either its time-independent or time-dependent forms; it is a product, rather, of the *actualization* of potentia, and therefore beyond the scope of the quantum formalism. Yet temporal and logical asymmetry *must* be a part of any coherent ontological interpretation of quantum mechanics. Heisenberg writes: "The transition from the 'possible' to the 'actual' is absolutely necessary here and cannot be omitted from the interpretation of quantum theory. At this point, quantum theory is intrinsically connected with thermodynamics in so far as every act of observation is by its very nature an irreversible process; it is only through such irreversible processes that the formalism of quantum theory can be consistently connected with actual events in space and time."[30]

The long-sought-after physical mechanism responsible for the actualization of potentia—that is, the existence of facts—remains elusive, however. We have already explored the reason proposed by a classical statistical interpretation: Asymmetrical causal relations such as those pertaining to thermodynamics are in fact purely symmetrical, time-reversal-invariant relations whose symmetry lies beneath a veil of hidden variables. In these hidden variables, in other words, lies the key to unscrambling an egg. "Irreversibility," wrote Born, "is a consequence of the explicit introduction of ignorance into the fundamental laws."[31]

But as already discussed, the inability of quantum mechanics to account for the actualization of potentia or the temporally asymmetrical relations that obtain from such actualizations, is not problematic given that quantum mechanics presupposes and anticipates the existence of facts; this is evinced in the concepts of state evolution, probability, and history. "One may consider," writes Roland Omnès, "that the inability of the quantum theory to offer an explanation, a mechanism, or a cause for actualization is in some sense a mark of its achievement. This is because it would otherwise reduce reality to bare mathematics and would correspondingly suppress the existence of time."[32]

The advantage of the decoherence-based interpretations of quantum mechanics, however, is that the process of negative selection entails the necessary integration of noninterfering potentia into sets of mutually exclusive and exhaustive, probability-valued potential outcomes. And without this process, it is unclear whether or not the actualization of potentia could occur at all. Omnès writes that though it is not the case with respect to the Schrödinger equation of the quantum formalism, the direction of time in a coherent, *logical interpretation* of quantum mechanics is "restricted by decoherence in such a way that the direction of time must be the same as the direction of time in thermodynamics. It should be observed that this statement is not an intrinsic necessity of pure quantum mechanics. If the universe contained only two or three particles [such that decoherence were not likely], there would be no such restriction and one would be allowed to choose arbitrarily the direction of time in logic."[33]

The actualization of potentia—the existence of facts—constitutes an asymmetrizing punctuation of the continuum of potentia whose superposed, symmetrical, and therefore "logically atemporal" relations are reflected in the logically atemporal dynamics of quantum mechanical state evolution and its time-independent Schrödinger equation. Even the time-dependent form of the Schrödinger equation is "logically atemporal," since as a linear, deterministic equation, it is symmetrical with respect to time-reversal. But it must be recalled that the Schrödinger equation is not "deterministic" in the sense that it is determinative of a unique, actual outcome state; it is determinative of probability-valued *potential* outcome states. Thus a time reversal transformation applied to the time-dependent Schrödinger equation does not yield a determined, unique state, but rather a superposition of potential states.

Apart from the actualization of potentia, which lies beyond the scope of the Schrödinger equation, a quantum mechanical evolution is, therefore, logically atemporal. Logical, asymmetrical temporality,

in other words, is not the "background" against which the Schrödinger equation, in either its time-dependent or more fundamental
time-independent forms, operates. Logical, asymmetrical temporality is a product of the actualization of potentia—the evolution of
probability to *fact*; it is not a metric by which the evolution of potentia to probabilities can be measured. For without the existence of
facts, there is nothing to measure temporally. Though an explanation for the existence of facts lies beyond the scope of quantum
mechanics, it is both presupposed and anticipated by the mechanics;
and indeed, the existence of facts is necessary if the mechanics is to
account for the physical, causal interrelations characteristic of our
experiences, and for the temporal asymmetry such interrelations entail. Omnès writes:

> The existence of actual facts can be added to the [quantum] theory
> from outside as a supplementary condition issued from empirical ob-
> servation. The structure of time must then be modified accordingly.
> Time must be split into a past, a present, and a future having very
> different qualities. Present and past are uniquely defined while the
> future must remain potential and subject to probabilistic expecta-
> tions. This structure of time, so obvious from the standpoint of obser-
> vation, turns out to be necessary from a theoretical standpoint.[34]

Temporal asymmetry guarantees the uniqueness of facts in this
ontological interpretation of quantum mechanics; yet it must be
stressed then that facts integrated into states and histories of states,
in a sense, evolve to become new facts with the creation of each new
state, in augmentation of the history to which the state applies. Past
actualities are real—and in a sense, then, immortal—given that the
actualization of any and every novel state of the universe, in augmentation of the history describing the universe, entails an integration of
all facts belonging to the universe. These many facts are integrated,
according to quantum mechanics, to form a single state; and the
history to which the state belongs is thus augmented. We make use
of past facts, then, via recollection, which is a novel—and, as are all
actualities, unique—re-creation of the past. All past actualities are
thus recorded in the actualization of every subsequent state—
recorded in a system, in a measuring apparatus, or in the environment. It is temporal asymmetry that relegates us to using perpetually
re-created records of past actualities rather than the actualities

themselves; but it is also temporal asymmetry that, by this same relegation, provides for the objective immortality of every fact in the universe. Omnès writes:

> Past facts are not absolutely real; they only *were* real. One can never indicate a past fact by pointing a finger at it and saying "that." One must call for a memory or use a record, a note, a photograph. Nevertheless, the derivation of a unique past is possible because quantum mechanics allows for the existence of memory and records. It goes without saying that everybody always takes this structure of time for granted but, up till then, it was always tacitly added to our understanding of physics and never supposed to be a consequence of it. . . . What we observe in reality is always something existing *right now*, even if we interpret it as a trace of an event in the past, whether it be a crater on the moon, the composition of a star atmosphere, or the compared amounts of uranium and lead in a rock.[35]

Summary

Now that we have described the quantum mechanical evolution of the state of a system—the forward, historical movement of facts from actuality to potentiality to actuality, driven by nothing other than their necessary mutual interrelations—let us formulate a final interpretation of this evolution as it applies to our idealized traffic signal system. It should first be noted that by most interpretations, quantum mechanics cannot be applied sensibly to the evolution of macroscopic systems; however, the decoherence effect, as we have seen, has served to bridge the conceptual border between "quantum reality" and "classical reality" as demarked by Bohr with his principle of complementarity. The "classicality" of macroscopic systems can now be interpreted most fundamentally as an effect of the mutual interrelations among the facts comprised by such systems. This is not to suggest that macroscopic systems are *best* described quantum mechanically; only that they are most fundamentally described by quantum mechanics.

Returning now to the interpretation of the evolution of our red-green traffic-signal system, let us define all the systems involved:

Universe
The system of all facts/actualities. We shall represent the state of this system—the maximal specification of all facts belonging to it—with

the state vector $|\Psi\rangle$. It is understood that individual states—that is, "snapshots" of any system—and serial histories of states, are mechanically equivalent, and for the purposes of this discussion are interchangeable and will both be represented by the same state vector.

Measured System

The system of all facts/actualities belonging to the traffic signal, whose state shall be represented by the state vectors $|u_{green}\rangle$ or $|u_{red}\rangle$. It is understood that the nomenclature used for the state vectors $|u_{green}\rangle$ or $|u_{red}\rangle$ indicate that these states only specify the facts pertaining to the status of the red and green lights; all other facts belonging to the Measured System are ignored.

Detector

The system of all facts/actualities belonging to the detector—an up/down pointer apparatus, whose state shall be represented by the state vectors $|d_\downarrow\rangle$ and $|d_\uparrow\rangle$. The detector is constructed so that when sufficiently interacted with the traffic-signal system, state $|d_\downarrow\rangle$ correlates with state $|u_{green}\rangle$, and state $|d_\uparrow\rangle$ correlates with state $|u_{red}\rangle$. All facts belonging to the Detector that do not pertain to the direction of the pointer are ignored.

Environment

The system of all facts/actualities other than those represented by $|u_{green}\rangle$, $|u_{red}\rangle$, $|d_\downarrow\rangle$ and $|d_\uparrow\rangle$. These facts may thus belong to the environment external to the Measured System and the Detector, or they may belong to the environment internal to them, comprising all facts not specified by $|u_{green}\rangle$, $|u_{red}\rangle$, $|d_\downarrow\rangle$, and $|d_\uparrow\rangle$—that is, facts that have been ignored. The state of the Environment is represented by the state vector $|E_0\rangle$.

It is understood that the facts belonging to the Measured System, Detector, and Environment also belong to $|\Psi\rangle$, forming "subsystems" of $|\Psi\rangle$, and their grouping into subsystems in no way vitiates the mutual and necessary interrelations with all facts belonging to $|\Psi\rangle$. Hence, the states of the Measured System, the Detector, and the Environment—as well as the histories of these states—are all objectively mutually correlated. An analogous example of such correlation can be seen when considering the local histories describing California and Maine, respectively, within the larger history of the

United States; facts belonging to each subhistory also belong to a larger history englobing both, which ensures an *objective correlation* among the facts specified by these histories. Similarly, one cannot exist in Paris and New York simultaneously, for the superposition of mutually interfering histories such a feat would require are eliminated via the decoherence effect.

The measurement of our traffic-signal system, then, most fundamentally entails a specification of a subset of facts belonging to the universe, relative to a single fact or subset of facts, referred to here as "the indexical eventuality" belonging to the subset of facts referred to as "the Detector." In an ontological interpretation of quantum mechanics, every fact must be understood as capable of fulfilling the role of "indexical eventuality" or "Detector" during the evolution of $|\Psi\rangle$, which, according to some interpretations, locks quantum mechanics into a fog of sheer subjectivity. But this condemnation overlooks the ontological requirement of the correlation among actualities (and histories of actualities) as discussed in the preceding paragraph, without which quantum mechanics—and indeed the world itself—would cease to be accessible to reason. Quantum mechanics is not, therefore, rooted in subjectivity, but rather merely in relativity; quantum mechanics does not entail the subjective relations among facts—merely the relations among facts relative to a given fact or subset of facts. Since the formalism requires that we follow only one indexical eventuality at a time, we shall select the pointer needle of our Detector, which was intended for the purpose.

We begin, then, with the mutually correlated states describing the System, the Detector, and the Environment—i.e., the Universe—whose state is described by $|\Psi\rangle$ and reflects the state of these systems relative to a single measurement interaction:

$$|\Psi\rangle = \alpha|u_{green}\rangle|d_\downarrow\rangle|E_\downarrow\rangle + \beta|u_{red}\rangle|d_\uparrow\rangle|E_\uparrow\rangle$$

This expression simply groups all possible projections of $|\Psi\rangle$ into two subspaces, $\mathscr{E}$ and $\mathscr{E}^\perp$, determined by the pointer needle of the Detector system, our choice of indexical eventuality. If subspace $\mathscr{E}_\downarrow$ consists of all vectors representing potential states (or potential histories) where the pointer is down $|d_\downarrow\rangle$, then subspace $\mathscr{E}_\downarrow^\perp$ consists of all vectors representing potential states (or potential histories) where the pointer is not down. ($E_\uparrow$, for example, would belong to this subspace.) The manifold possible projections of $|\Psi\rangle$ upon $\mathscr{E}$ are integrated into the term

$$\alpha|u_{green}\rangle|d_\downarrow\rangle|E_\downarrow\rangle$$

interpreted as: "All potential states of the universe where the pointer is down and the light is green, with all other actualities unspecified." Every potential state or history of states integrated in $\alpha|u_{green}\rangle|d_\downarrow\rangle|E_\downarrow\rangle$ thus tells *some* unique, potential story about every actuality in the universe, all of which agree upon the specification of the color of the traffic signal and the direction of the pointer needle. The complex coefficient α represents the length of this integrated projection and will ultimately represent the probability that the traffic signal will be green and the pointer needle of the detector will point downward— that is, the quantified valuation of this potential fact.

Likewise, the possible projections of $|\Psi\rangle$ upon $\mathscr{E}_\uparrow$ are integrated into the term $\beta|u_{red}\rangle|d_\uparrow\rangle|E_\uparrow\rangle$, with the complex coefficient β representing the length of this integrated projection, and ultimately the probability that the traffic signal will be red and the pointer needle of the detector will point upward.

However,

$$|\Psi\rangle = \alpha|u_{green}\rangle|d_\downarrow\rangle|E_\downarrow\rangle + \beta|u_{red}\rangle|d_\uparrow\rangle|E_\uparrow\rangle$$

also includes nonsensical, mutually interfering superpositions of states whose vectors belong to neither $\mathscr{E}_\downarrow$ nor $\mathscr{E}_\uparrow$ exclusively. These are revealed in the correlated, pure state density matrix of potential states of $|\Psi\rangle$:

$$
\begin{aligned}
\rho^c &= |\Psi\rangle\langle\Psi| \\
&= |\alpha|^2|u_{green}\rangle\langle u_{green}||d_\downarrow\rangle\langle d_\downarrow||E_\downarrow\rangle\langle E_\downarrow| \\
&+ \boldsymbol{\alpha\beta^*|u_{green}\rangle\langle u_{red}||d_\downarrow\rangle\langle d_\uparrow||E_\downarrow\rangle\langle E_\uparrow|} \\
&+ \boldsymbol{\alpha^*\beta|u_{red}\rangle\langle u_{green}||d_\uparrow\rangle\langle d_\downarrow\rangle\langle E_\uparrow|\langle E_\downarrow} \\
&+ |\beta|^2|u_{red}\rangle\langle u_{red}||d_\uparrow\rangle\langle d_\uparrow||E_\uparrow\rangle\langle E_\uparrow|
\end{aligned}
$$

where the off-diagonal terms in boldface represent superpositions of interfering and thus nonsensical states incapable of integration.

The correlation between our System and Detector and the Environment produces such a vast multiplicity of potential states, each represented by a unique projection in $\mathscr{E}_\downarrow$ or $\mathscr{E}_\uparrow$, that the majority of these vectors cancel each other out, thereby eliminating the interfering superpositions represented by the off-diagonal terms in boldface. This cancellation is represented mathematically by a "trace-over" or "sum-over" of the unmeasured degrees of freedom belonging to the Environment. Eliminating these unmeasured, potential environmental facts, in essence, eliminates the superpositions of incompatible, potential universal states associated with them. The result of this process of negative selection is the decoherence of the potential outcome states, such that those that remain are mutually exclusive and exhaustive potentia (in satisfaction of the logical principles of noncontradiction and the excluded middle), each valuated as a probability. These are represented by a reduced density matrix:

$$\rho^r = |\alpha|^2 |u_{green}\rangle\langle u_{green}||d_\downarrow\rangle\langle d_\downarrow| + |\beta|^2 |u_{red}\rangle\langle u_{red}||d_\uparrow\rangle\langle d_\uparrow|$$

where $|\alpha|^2 + |\beta|^2 = 1$, such that each term represents a potential state or history of states valuated as a probability.

We can now briefly summarize the ontological interpretation of the quantum mechanical interaction described above as follows:

1. The specification of the state of a system of facts necessarily entails the state of the universe which subsumes the system. Further, the specification of the state of a system is itself a fact—a novel fact, born of the process by which the state is specified. Thus, the specification of the state of a system ultimately entails the specification of an historical route of states, newly augmented with each new specification. Every specification of a system state, then, entails the evolution of a state from a multiplicity of actualities, through integrations of potentialities, to a novel actuality.

2. The specification of the state of a system of facts begins with the integration of all facts relative to a particular fact referred to as the "indexical eventuality." The relativity of this integration is both objective and subjective. It is objective in that it is an integration of facts, and it is subjective in that the uniqueness of the indexical eventuality conditions the particular form of the integration. The latter is reflected in quantum mechanics by the fact that the preferred basis is determined by the indexical eventuality.

3. There are manifold potential integrations of potential facts relative to any particular indexical eventuality, and each potential integration amounts to a potential state (or potential historical route of states) of the system. Since each indexical eventuality is unique, each produces its own particular forms of these integrations, such that for

each indexical eventuality, there exist potential facts which are incapable of integration in a single potential state. If, for example, there are two potential facts pertaining to a particle's position and these potential facts disagree, they cannot be integrated together in a "consistent" potential state or history of states. In quantum mechanics, these mutually interfering potential facts or potential states are bound into coherent superpositions, represented in the formalism by the off-diagonal terms in the pure state density matrix.

4. A process of negative selection is therefore required to remove these mutually interfering potentia which are incapable of integration. This process is possible only by virtue of the integration of *all* facts, and not only those belonging to the measured system; for it is the correlation of facts belonging to this system with unmeasured and relatively irrelevant facts environmental to it that produces a sufficient number of potential states that (i) particular forms of integration with compatibilities and incompatibilities are possible at all and (ii) there exist a sufficient number of incompatible potential states that their propensities for actualization are mutually cancelled. The elimination, by negative selection, of superpositions of interfering, incompatible potentia is achieved by ignoring all potential facts that are essentially unrelated to the pointer basis of the indexical eventuality (the "unmeasured degrees of freedom" of the environment). In other words, the elimination of this irrelevant multiplicity of detail of environmental potential facts effectively entails the elimination of any interfering, coherent superpositions of potential states associated with them. The result is a "decoherent" set of noninterfering potential states, represented in the formalism by the reduced density matrix, that are (i) logical, in that they are mutually exclusive and exhaustive, in satisfaction of the principles of noncontradiction and the excluded middle, and (ii) have propensities for actualization that have become valuated as probabilities.

5. This entire process is predicated upon the a priori existence of facts, as implied, for example, in the following three concepts: (i) state *evolution* (evolution from antecedent fact, through potentia, to novel fact); (ii) *histories* of states (histories being serial routes of facts perpetually augmented by novel facts); and (iii) *probability* (probability that an antecedent fact will become a particular potential novel fact according to a quantifiable valuation). It is the probability valuation of the mutually exclusive and exhaustive potential

states that governs, and therefore guarantees, the actualization of a unique outcome state. The entire process, therefore, occurs relative to both (*i*) the indexical eventuality as an *actual subject*, participant in the system whose state is being specified, and (*ii*) the indexical eventuality as the *actual satisfaction* of the valuations terminal of that part of the process described by quantum mechanics. For as probabilities, these valuations presuppose and anticipate this satisfaction. It is the valuation of potentia, then, and the process productive of this valuation—rather than the actualization of potentia—that quantum mechanics describes.

NOTES

1. Werner Heisenberg, *Physics and Philosophy* (New York: Harper Torchbooks, 1958), 185.

2. Murray Gell-Mann, *The Quark and the Jaguar: Adventures in the Simple and the Complex* (New York: W. H. Freeman, 1994), 167.

3. Heisenberg, *Physics and Philosophy*, 143.

4. Ibid., 55.

5. Ibid., 53.

6. Wojciech Żurek, "Decoherence and the Transition from the Quantum to the Classical." *Physics Today* 44, no. 10 (1991): 36–44.

7. Wojciech Żurek, "Letters," *Physics Today* 46, no. 4 (1993): 84.

8. John von Neumann, *Mathematical Foundations of Quantum Mechanics* (Princeton, N.J.: Princeton University Press, 1955).

9. Żurek, "Decoherence and the Transition from the Quantum to the Classical," 39.

10. Ibid., 41.

11. Ibid., 40.

12. Robert J. Griffiths, "Consistent Histories and the Interpretation of Quantum Mechanics," *Stat Phys.* 36 (1984): 219; see also Griffiths, *Consistent Quantum Theory* (Cambridge: Cambridge University Press, 2002).

13. Roland Omnès, *The Interpretation of Quantum Mechanics* (Princeton, N.J.: Princeton University Press, 1994).

14. Murray Gell-Mann and James Hartle, "Strong Decoherence," in *Proceedings of the Fourth Drexel Symposium on Quantum Non-integrability— The Quantum-Classical Correspondence*, Drexel University, September 1994.

15. The "discrete histories" interpretation advocated by Griffiths, Gell-Mann, Hartle, and Omnès entails serially ordered sequences of quantum

mechanical states—that is, time sequences of projection operators. This idea is related to an earlier conception of continuous quantum mechanical histories by Richard Feynman. But Feynman's histories ("Feynman paths") are "continuous histories" rather than "discrete," decoherent histories.

16. Gell-Mann and Hartle, "Strong Decoherence."

17. Gell-Mann, *The Quark and the Jaguar*, 153.

18. John Bell, "On the Einstein Podolsky Rosen Paradox," *Physics* 1, no. 3 (1964): 195–200.

19. S. Freedman and J. Clauser, "Experimental Test of Local Hidden Variables Theories," *Phys. Rev. Lett.* 28 (1972): 934–941.

20. A. Aspect, J. Dalibard, and G. Roger, "Experimental Test of Bell's Inequalities Using Time-Varying Analyzers," *Phys. Rev. Lett.* 44 (1982): 1804–1807.

21. Max Born, "Bemerkungen zur statistischen Deutung der Quantenmechanik" in *Werner Heisenberg und die Physik unserer Zeit*, ed. F. Bopp (Braunschweig: F. Vieweg, 1961), 103–118.

22. David Bohm, "A Suggested Interpretation of the Quantum Theory in Terms of 'Hidden' Variables," *Phys. Rev.* 85 (1952): 166, 180. See also David Bohm and B. J. Hiley, *The Undivided Universe: An Ontological Interpretation of Quantum Theory* (London: Routledge, 1993).

23. David Bohm and B. J. Hiley, *The Undivided Universe: An Ontological Interpretation of Quantum Theory* (London: Routledge, 1993), 358.

24. Ibid., 359.

25. Ibid., 283–284.

26. P. T. Geach, *God and the Soul* (London: Routledge, 1969).

27. Ibid., 71–72.

28. It is possible, in fact, to relate this temporal symmetry-asymmetry to the spatial symmetry-asymmetry of quantum nonlocality discussed previously. Where classical dynamics requires the symmetrical, mutual independence of spatially well separated systems—that is, that they be symmetrically externally related only—quantum dynamics allows for nonlocal, asymmetrical interrelations between spatially well separated systems. Thus, an actualization of a potential state in one system, as we have seen, affects the *potential* states in the other system, thereby affecting the *potential* subsequent actualizations of this system.

29. Heisenberg, *Physics and Philosophy*, 52.

30. Ibid., 137–138.

31. Max Born, *Natural Philosophy of Cause and Chance* (Oxford: Oxford University Press, 1949), 72.

32. Omnès, *The Interpretation of Quantum Mechanics*, 494.

33. Ibid., 318.

34. Ibid., 508.

35. Ibid., 344.

Interlude

The Philosophy of
Alfred North Whitehead

ALFRED NORTH WHITEHEAD was as much a mathematician as he was a metaphysician, and so it should come as no surprise that whereas some have characterized his metaphysical system as intuitive, others have found it to be frustratingly complicated and difficult to understand. If mathematics is indeed intuitively simple at some level, it is because mathematics always abides by the fundamental desiderata of logic, coherence, applicability to human experience, and adequacy in that applicability such that one cannot conceive of a type of experience where mathematics is fundamentally inoperative. These are, of course, the same four desiderata Whitehead assigns to his metaphysics, and the manner in which they are fulfilled by the latter is similar to the manner they are fulfilled by mathematics—a manner demonstrative of a deep complexity but married to a fundamental, intuitive simplicity. This is not to suggest that Whitehead proposed a metaphysical scheme intended to be wholly definable by mathematics (and by implication reducible to mathematics); the significance of these shared desiderata is, rather, the implication that Whitehead's metaphysical scheme might at some level be describable by mathematics.

For both mathematics and Whitehead's metaphysics, the requirement of empirical adequacy is particularly evident in the case of theoretical physics, for which Whitehead's metaphysics was intended to supply a suitable ontological framework. In this sense, the applicability and adequacy of any speculative metaphysical framework intended to accommodate physical phenomena is, for Whitehead, measured in part by its empirical exemplification. The traditional philosophical opposition of the terms "empirical" and "metaphysical," then, is a dualism Whitehead would likely correlate with the Cartesian dualism of matter and mind, the repudiation of which lies at the root of his philosophy. The traditional philosophical notion of "empirical" as understood to mean based upon observation or experience alone without regard for system and theory, then, is not the notion Whitehead intends in his use of the term. Rather, the empirical side of his metaphysical scheme as expressed in the desiderata "applicability" and "adequacy" is to be thought of as the bridge by which the principles of Whitehead's metaphysics are connected with human experience. The soundness of the underlying framework, then, derives not only from the logical and coherent applicability of the metaphysics to distinct, often exclusive realms such as those that define physical and microphysical experiences, but rather from its logical and coherent applicability to human experience itself.

With each passing year, modern physics becomes increasingly relevant to our everyday lives. Breakthroughs in cosmology continually make their way into the morning newspapers, and breakthroughs in technology continually make their way onto our desktops, into our living rooms, into us; and in all of these cases, quantum mechanics has become more and more prominent with each leap forward, as has the need for an ontology capable of accommodating the quantum theory logically, coherently, and adequately. Since his death in 1947, the influence of Whitehead's metaphysics has grown steadily, if not rapidly. And in the coming years, as quantum mechanical phenomena grow to become the very heart of our everyday technology in the form of quantum transistors, superconducting devices, and quantum computers, the popularity of Whitehead's philosophy is likely to undergo a rapid expansion. For the regnant classical mechanical worldview is simply incapable of accommodating quantum mechanical phenomena without glaring paradox, ontological inconsistency, and arbitrary dispensations from important classical mechanical laws and principles. And as these quantum mechanical phenomena grow to become more and more integral to our lives, these paradoxes and inconsistencies will become less and less tolerable.

One goal of the present work is to demonstrate how Whiteheadian metaphysics can be heuristically useful in understanding modern ontological interpretations of quantum mechanics, such that the physics can be interpreted logically, coherently, and empirically adequately as an exemplification of the metaphysics; but just as important is the converse demonstration that modern ontological interpretations of quantum mechanics can be heuristically useful to an understanding of Whiteheadian metaphysics. *Process and Reality*, the opus in which Whitehead's metaphysical system is given in its most complete and systematic form, presents this system in an infamously nonlinear format, wherein the entire scheme is essentially presupposed with each elucidation of a particular aspect. In this sense, rather than proceeding in linear fashion from beginning to middle to end, with each part presupposing its antecedents, *Process and Reality* proceeds in an almost inward-spiraling fashion, each revolution presupposing the overall curvature, with repeated visitation of each quadrant along the way to an ever-retreating center. By mapping a linear treatment of quantum mechanics onto this nonlinear treatment given in *Process and Reality*, I hope to make each treatment mutually illuminative of the other.

For indeed, the overall theme driving Whitehead's metaphysical scheme is the same theme driving modern ontological interpretations of quantum mechanics as discussed in Part I—the repudiation of fundamental mechanistic materialism and the redefinition of such materialism as a mathematical abstraction that ought not be mistaken for a fundamental description of the "concrete" reality of nature. Whitehead refers to this as the "fallacy of misplaced concreteness," and, as discussed in Part I, Heisenberg held a similar view. For Whitehead, classically described objects are more fundamentally described as historical routes of atomic events, where past events influence but do not determine future events. The universe is a multiplicity of such events, each of which evolves or becomes via a process of prehending and integrating all the antecedently actualized events (data) that the universe comprises. Some data are, of course, more relevant than other data; and indeed, most data once brought together by prehension are further integrated largely by elimination. Conceptually, such integration of data through elimination can be thought of in the same sense that mathematical terms brought together in an equation are eliminated through cancellation. The function and importance of these terms as constituents in the equation is in no way vitiated by their cancellation; cancellation is simply the proper mode of their integration with the other terms of the equation.

Most significantly, however, the data prehended, while objectively real, can be objectified by an occasion in any number of potential ways. Data can be objectified by simple reproduction, for example, and Whitehead's Category of Conceptual Reproduction lies at the heart of what we perceive to be the "enduring objects" that dominate our classical worldview—our conceptions of atoms, molecules, rocks, planets, suns; and sometimes the reproductions take a more rhythmic form characteristic of electromagnetic waves, probability functions, and so forth. The Category of Conceptual Reproduction as it applies to macroscopic "enduring objects" is closely associated with Whitehead's Category of Transmutation, whereby manifold microcosmic prehensions of data are "transmuted" into a single, macrocosmic perception of an integrated datum or "collective observable" in the language of quantum physics—a process analogous to looking at a photo in the newspaper and seeing a single image rather than a multiplicity of individual dots.

But data are not always objectified by simple reproduction, nor are their transmutations necessarily inherited from and consistent with antecedent transmutations. Data are also integrated according to the Category of Conceptual Reversion—that is, according to novel potential forms and transmutations that were not simply inherited from the historical route, but instead ingressed into the becoming occasion from somewhere else—from some other actuality apart from that particular historical route. In this way, each occasion, and the societies they form, has the potential for novel growth. Even the most rudimentary electromagnetic occasions enjoy such reversions from time to time, and this is evinced, for example, by indeterministic quantum mechanical phenomena.

For Whitehead, the potentia driving novelty constituted a different species of reality, as they did for Heisenberg—realities that do not derive entirely from some particular antecedent actual datum but rather from a spatiotemporally generic, and therefore primordial, actuality. The fact of any possibility, in other words, necessarily derives from a more fundamental actuality, and this reasoning requires the concept of a supremely fundamental, primordial actuality. In Kant's 1762 work *The Only Possible Ground for a Demonstration of God's Existence*, he argues that this reasoning from *possibility as a consequence* to God's existence as the *ground of this possibility* is the only sound demonstration of the existence of God. And for Whitehead, this same reasoning is central to his own conception of God's necessary existence and relations with the world.

Whitehead was indeed an atomist, then—a realist, but not in the materialist sense; for his atomic actualities are not substances, indivisible and symmetrically interrelated such that their interactions are strictly deterministic and time-reversal invariant (i.e., actualities formative of a clockwork universe); Whitehead's atomic actualities are, rather, individual occasions, asymmetrically related (past occasions being settled, future occasions being open) such that each new occasion embodies a common past and, once actualized, contributes to this past, recreating it by its augmentation of it. The many occasions of the past become one in each new occasion, and are thus increased by one.

Central to Whitehead's atomism, however, is a repudiation not just of fundamental mechanistic materialism, but also of the Cartesian "bifurcation of nature" that typically accompanies it. Mentality and materiality are mutually implicative, interrelated modes of reality for Whitehead, not separate species of reality, one more fundamental than the other. For Whitehead, each atomic occasion is dipolar, with (i) a physical pole that entails the actual occasion's relationship with its antecedent data that are thereby causally efficacious in its becoming—that is, its "public," real physical relations with its universe; and (ii) a mental pole, which entails the actual occasion's evolving forms of definiteness—the "private" workings of reproduction, reversion, transmutation, and other categories that describe the evolution of the occasion from potentiality to actuality (the term "mental pole" should not be misunderstood to imply conscious mentality, however, which in Whitehead's metaphysics is a higher-order function inoperative in the vast majority of actual occasions). For Whitehead, each pole is incapable of abstraction from the other, in the same way that the concept of potentiality is incapable of abstraction from the concept of actuality. The traditional philosophical distinction between primary and secondary qualities of an object is thus replaced by a more subtle and complex scheme revolving around these two interrelated poles of the atomic actual occasion.

For Whitehead, then, these dipolar entities, their relations and their actualizations, are typically analyzed in one of two ways: One way is by "coordinate analysis," where emphasis is given to the physical pole for which relations among occasions are primarily relations of causal efficacy. Coordinate division gives emphasis to the physical pole in the sense that it is in the physical pole that the data of the actual world are initially appropriated according to their nature as concrete, spatiotemporally "coordinated" quanta. Such relations are well described by classical physics, chemistry, biology, and other sciences; coordinations of data according to special relativity, for example, have significance only in the physical pole.

The second type of analysis is "genetic analysis," where emphasis is given to the mental pole. The spatiotemporal coordination of data in the physical pole often manifests as nexūs and societies of occasions whose loci and other defining characteristics ("congenial uniformities") are vague and ill defined. "Presentational immediacy" in the mental pole contributes a precision to such nexūs and societies—a precision perceived as the occasion's "contemporary world," which is inoperative in the physical pole because of the limiting spatiotemporal coordination of the data prehended in that pole. (Special relativity, for example, requires that data prehended in the physical pole lie in the subject occasion's past light cone. Thus, contemporary occasions cannot be mutually causally efficacious.) Presentational immediacy presupposes the causally efficacious relations of the physical pole, and projects upon them a sharp, well-defined "contemporary" state—that is, one of the mutually exclusive and exhaustive precisely defined alternative system states described in quantum mechanics, among which one will become actual. In the mental pole, a prehending subject's "presented locus" (or "strain-locus") is the subjectively "immediate" spatial, geometrical coordination of its actual world in terms of subjectively "contemporaneous" temporally coordinated actualities organized into a "presented duration"—a contemporary nexus, perceived in the mode of presentational immediacy.

Whereas data are coordinated in the physical pole, they are integrated in the mental pole, and the integration entails reproductions, reversions, negative prehensions, and transmutations, among other processes, which project sharp, well-defined forms of definiteness upon the data inherited from the physical pole. The result is a matrix of alternative, valuated potential forms of definiteness. Whitehead's genetic analysis, then, is the type of analysis with which quantum mechanics is primarily concerned; for quantum mechanics describes the evolution of a system of actual occasions from an initial state to a final state—an evolution that entails the integration of antecedent data according to a matrix of potential forms of definiteness. And just as the physical and mental poles are mutually implicative, so are coordinate and genetic analyses. This mutually implicative dipolarity is evinced in quantum mechanics by the problem of state reduction, which is easily solved when one acknowledges that the actualization of potentia is presupposed by quantum mechanics; the physical pole is presupposed by the mental pole, and therefore cannot be accounted for by it. Potentiality, in other words, is nonsensical apart from presupposed actuality. At the same time, however, the dipolarity of concrescence entails that the spatiotemporal coordination operative in the physical pole cannot occur apart from the presupposed genetic operations in the mental pole of a prior occasion. For apart from these operations, the prior occasion would not exist, and there would be no data to coordinate. Thus, in Whitehead's dipolar metaphysical scheme, coordination presupposes genesis as much as genesis presupposes coordination. In the same way, a quantum mechanical state evolution presupposes logically and temporally prior actualized evolutions as data (that is, there must be an initial state that evolves); and the quantum mechanical actualization of a potential outcome state (the final state) presupposes the evolution from whence it came—and also a subsequent evolution for which it will serve as datum. (Recall that in quantum mechanics, all outcome states are necessarily confirmed retrodictively, via a subsequent measurement or state evolution.)

For Whitehead, both genetic and coordinate analyses are governed by the cooperation of two fundamental principles: the Principle of Relativity, according to which "the potentiality for being an element in a real concrescence of many entities into one actuality is the one general metaphysical character attaching to all entities . . . [such that] it belongs to the nature of a 'being' that it is a potential for every 'becoming'";[1] and the Ontological Principle (or "Principle of Efficient and Final Causation"), according to which "every condition to which the process of becoming conforms in any particular instance has its reason *either* in the character of some actual entity in the actual world of that concrescence, *or* in the character of the subject which is in process of concrescence."[2]

According to the Ontological Principle, writes Whitehead, "there is nothing which floats into the world from nowhere."[3] Coordinate analysis of any actual occasion has as its object the conditioning influences derived from the spatiotemporally coordinated antecedent actualities of its actual world; and genetic analysis has as its object the conditioning influences derived from the "nonactualized" yet real world of potentia. These nonactual potentia revealed by genetic analysis are the formative elements of the actual, temporal world, and apart from their participation in the actual world, we would know nothing of them. Further, the genetic analysis of an actual occasion reveals two important implications with respect to how nonactual yet ontologically significant "real" potentia contribute to the formation of any actual occasion: first, that the actualization of potentia is a creative process, such that the actual world is most accurately seen as an historical route of creations whose logically ordered relations reveal an overall trajectory of ongoing novelty—of creative advance; second, that in a genetic analysis of any particular occasion one can trace the lineage of particular constituent potentia back through a particular logically ordered, historical route of occasions.

But one is also able to discern potentia that are not derivable in this way, and indeed, it is these "pure" potentia that drive the creative advance, lest the novelty of the future be reducible to the possibilities of the past. And yet by the Ontological Principle, these "pure potentia" must derive from somewhere—some actual yet nonhistorical, or better, nontemporal, primordial source. This primordial actual entity is God in Whitehead's philosophy—the metaphysical source of pure potentiality and true novelty in the universe. This novelty is continually manifested in the creative advance of the world—an advance that is both indeterminate, yet by the Ontological Principle, conditioned by the actual world temporally (via historically derivable "real" potentia) and nontemporally (via the "pure" potentia that drive true novelty, originating in God).

Thus with the Ontological Principle and the Principle of Relativity, the valuable concepts given by postmodern subjectivism (those that emphasize the private, autonomous, and creative aspects of existence) and the valuable concepts of classical realism (those that emphasize the public, heteronymous relations with an objectively real universe) are brought together in the dipolar unification of the mental and physical poles in each atomic actual occasion. Each actuality is thus creative of itself, but based in large part upon the real data of the universe, as well as the generic pure potentia or "eternal objects" that ingress into the concrescence—potentia that must derive from God, the supremely fundamental, primordial actuality.

Whitehead called his metaphysical scheme the "philosophy of organism," and this, coupled with the operation of the mental pole in even the most basic electromagnetic occasions, might lead one to believe that Whitehead considered the universe to be "alive" or held that objects such as stones or trees enjoy the same kind of mentality that human beings enjoy. Both of these are misapprehensions of Whitehead's metaphysics. For although every occasion entails a mental pole, "mentality" in this sense is not synonymous with consciousness, intellectually informed free will, or mind. There are "low-grade" occasions such as electromagnetic ones for which the operations of the mental pole are limited solely to reproduction of antecedent data, with rare, rudimentary reversions and transmutations restricted to the integration of potentialities—integrations of the sort described by quantum mechanical indeterminacy, for example. For these occasions, the conformal, causal operations of the physical pole dominate, and the rudimentary conceptual reproductions, reversions, and transmutations of the mental pole constitute the whole of their "mentality."

In contrast, "high-grade" occasions, such as those associated with the human mind, entail an enhanced mental pole where advanced conceptual reversions and transmutations may be just as operative, if not more so, than the mere reproduction of antecedent data. In the actual occasions of the human mind, conceptual reversions and transmutations transcend the form of mere "potential fact" (mere eigenstate in quantum mechanics) and take the more complex and intense forms of proposition, of hypothesis, of imagination, of dream.

Whitehead's distinction between "high-grade" and "low-grade" occasions in terms of the operations of the mental pole is particularly important in qualifying the "aliveness" of occasions and—more significantly as regards modern complexity theory—societies of occasions. For Whitehead, the lower-grade occasions constitute the more fundamental species of "physical purpose" that forms the basis of microphysics—occasions related to transfers of electromagnetic energy and the like. This species of occasion, rather than creating by "private experience," instead "receives the physical feelings, conforming their valuations according to the [public] 'order' of that epoch. . . . [Their] own flash of autonomous individual experience is negligible."[4] Our cosmic epoch consists primarily of these low-grade electromagnetic occasions,[5] which form structured societies. These structured societies subsume both (i) subordinate societies, whose definition and integrity are largely independent of their environments (i.e., the molecules of a cell); and (ii) subordinate nexūs, whose definition depends largely on their environments (the cytoplasm of a cell, the organs of the human body, and so on). The former are thus typically less "specialized" than the latter, and in general, "a structured society may be more or less 'complex' in respect to the multiplicity of its associated sub-societies and sub-nexūs and to the intricacy of their structural pattern."[6]

So "the problem for Nature," writes Whitehead, "is the production of societies which are 'structured' with a high 'complexity,' and which are at the same time 'unspecialized.' In this way, intensity is mated with survival."[7] This is accomplished in two ways for Whitehead: The first way is via the "lower-grade" physical purposes characteristic of quantum mechanical evolutions where conceptual reproduction is primary, and the operation of the mental pole is mainly limited to negative selection associated with decoherence—that is, the massive average objectification of a nexus, and the elimination of detailed diversities from it. The result is the "low-grade" structured societies such as crystals, rocks, planets, and suns, the "most long lived of the structured societies known to us."[8]

The second way nature accomplishes unspecialized complexity is "an initiative in conceptual prehensions, i.e., in appetition. . . . In the case of the higher organisms, this conceptual initiative amounts to thinking about the diverse experiences. . . . This second mode of solution also presupposes the former mode. . . . Structured societies in which the second mode of solution has importance are termed 'living.' A structured society in which the second mode is unimportant, and the first mode is important, will be termed 'inorganic.'"

Whitehead continues:

> In accordance with this doctrine of "life," the primary meaning of "life" is the origination of conceptual novelty—novelty of appetition. Such origination can only occur with the Category of Reversion. Thus a society is only to be termed "living" in a derivative sense. A "living society" is one which includes some "living occasions." Thus a society may be more or less "living" according to the prevalence in it of living occasions. Also an occasion may be more or less living according to the relative importance of the novel factors in its final satisfaction.[9]

The universe, then, is for Whitehead most fundamentally a structured society of electromagnetic occasions that contains both subordinate societies and subordinate nexūs, the vast majority of which are "inorganic," "nonliving" societies whose mentality is limited to conceptual reproduction and rudimentary transmutation and reversion. Hence, for Whitehead, the universe for the most part is not "alive." Clearly, then, without proper attention to the Whiteheadian distinctions between organic and inorganic, living and nonliving, conscious and merely mental, the correlation of Whitehead's metaphysics with the physical sciences will be a needlessly uneasy one, likely to inspire either the complete excising of the mental pole on the grounds that it cannot be relevant to physics—or even worse, attempts to use quantum mechanics to "explain away" the human mind and other higher-order mentality.

Either would be terribly unfortunate. For Whitehead's drive was never to explain away existence, nor was it to define one type of experience solely in terms of another more fundamental type. His drive was instead to show how traditionally incompatible areas of inquiry such as modern physics, philosophy, and even religion, could be brought together in a mutually illuminative way within the framework of a logical, coherent, empirically applicable, and empirically adequate metaphysical scheme. Existence was not to be explained away; it was to be enjoyed through adventures in understanding. But for Whitehead, the first step in a proper understanding, apart from embracing the four desiderata just mentioned, was to set an asymptotic course instead of the steep, head-on trajectory typical of both philosophy and science throughout history. Whitehead saw the advent of quantum mechanics, and its characterization by its innovators as *die endgültige Physik*; and it was most certainly not lost upon him that Newton had taken his own physics to be the final word as well. For Whitehead, philosophical and scientific dogmatism would lead to nothing but intellectual death from an unchecked craving for a head-on crash into Truth. Better to glide down in a gentle curve, satisfied to skim the surface and enjoy the view.

NOTES

1. Alfred North Whitehead, *Process and Reality: An Essay in Cosmology, Corrected Edition*, ed. D. Griffin and D. Sherburne (New York: Free Press, 1978), 22.

2. Ibid., 24.
3. Ibid., 244.
4. Ibid., 245.
5. Ibid., 98.
6. Ibid., 100.
7. Ibid., 101.
8. Ibid., 102.
9. Ibid., 102.

II

Quantum Mechanics and Whitehead's Metaphysical Scheme

4

The Correlation of Quantum Mechanics and Whitehead's Philosophy

> It is a remarkable characteristic of the history of thought that branches of mathematics, developed under the pure imaginative impulse, thus controlled, finally receive their important application. Time may be wanted. Conic sections had to wait for 1800 years. In more recent years, the theory of probability, the theory of tensors, the theory of matrices are cases in point.
>
> Alfred North Whitehead, *Process and Reality*

THE ONTOLOGICAL INTERPRETATION of quantum mechanics explored thus far in this essay can be distinguished from other interpretations by two primary characteristics: First, it is an interpretation that attempts to describe, via imaginative hypothetical deduction, the form among the facts of experience, rather than both the form *and* the facts, as is the case with many other interpretations of quantum mechanics—those, for example, that attempt to account for the actualization of potentia by way of the mechanics. The existence of facts by this interpretation is accepted a priori, such that the mechanics and its interpretation both presuppose and anticipate the facts of actuality as described in the quantum mechanical evolution of system states. The "problem of the actualization of potentia" is thus no problem at all by this interpretation, which merely seeks to describe the underlying form, and implications, of the quantum mechanical process by which actuality evolves to actuality, mediated by given potentia.

Second, it is an interpretation that, via the decoherence effect and its required mutual interrelations between the measured system and all facts environmental to it, explicitly makes use of the requirements of logic, coherence, and universal applicability and adequacy typically lacking in most interpretations of quantum mechanics. For these desiderata are satisfied with the explicit recognition of the universe in its entirety as the only truly closed system to which quantum mechanics might apply, and it is this recognition that guarantees the necessary, mutual interrelations among all facts. The implications of quantum mechanics by this interpretation are thus universal and therefore ontologically significant; and indeed, this interpretation of quantum mechanics constitutes the exemplification of a clear ontological principle, rather than merely an epistemological principle such as Bohr's principle of complementarity. The ontological principle is: Every fact is a determinant in the becoming of every new fact, such that the evolution of any fact entails both temporally prior facts and logically prior potentia as data, and an integration of these data that is unique to that evolution.

The interpretation of quantum mechanics described in Part I is a fundamental physical exemplification of this ontological principle; and given that, one might infer that it is an exemplification of some much broader metaphysical scheme that must flow from this principle. It would have to be a scheme wherein the universe is characterized as an ongoing process of actualizations (described by quantum mechanics as an historical route of state evolutions of $|\Psi\rangle$). Each actualization is itself a process, comprising the following phases as exemplified by quantum mechanics:

(i) An initial phase, consisting of the integration of all facts relative to a particular fact belonging to a particular subsystem of facts (e.g., the "indexical eventuality" belonging to the measuring apparatus) into potential forms or states. Since the process of actualization is described mechanically as an "evolution" of the state of the system of facts relative to a particular fact, the latter must therefore have two natures: (*a*) that of the *subject* of the state evolution, partially characterized by its inclusion in and relation to S_{init} (e.g., in quantum mechanics, the system state always evolves relative to the indexical eventuality and its associated preferred basis); (*b*) that of the product of the state evolution, partially characterized by its inclusion in and relation to S_{final}—a novel integration of facts which includes the newly evolved indexical eventuality. The relativity of this integration is both objective and subjective. It is objective in that it is an integration of facts, and it is subjective in that the uniqueness of the indexical eventuality, together with the objective actuality of the facts interrelated, conditions the particular form of the integration. The latter is reflected in quantum mechanics by the fact that a preferred basis is always associated with any indexical eventuality.

(ii) A supplementary phase, whereby potential facts incapable of integration are eliminated via negative selection, yielding a reduced matrix of mutually exclusive, valuated potential integrations—that is, a matrix of potential states or potential forms of facts, valuated as probabilities. These forms are subjective insofar as they are integrations of facts relative to a unique indexical eventuality; but they are also objective insofar as they are integrations of facts.

(iii) The actualization of one of these integrations according to the valuations qualifying each, in satisfaction of the evolution and its aim—the latter as evinced by the probabilistic nature of these valuations.

Readers familiar with the cosmological scheme developed by Alfred North Whitehead in his "philosophy of organism" have likely already inferred a number of correlations between the Whiteheadian scheme and the interpretation of quantum mechanics described in Part I. The explication of these correlations is the task of the remainder of this book.

It should be noted that the development from 1924 to 1930 of the "new" quantum theory of Heisenberg, Bohr, Schrödinger, et al., and its philosophically troublesome innovations—many of them hashed out at the Solvay Conferences of 1927 and 1930—took place during the same years that Whitehead developed his cosmological scheme, presented in its most complete, systematic form—*Process and Reality*—in his 1927–1928 Gifford Lectures, published in 1929. (Whitehead had presented an earlier version of this scheme in the 1925 Lowell Lectures, as well as in *Science and the Modern World*, published the same year.) One is left to wonder, then, whether Whitehead was aware of the troubling philosophical implications of the "new" quantum mechanics—as opposed to the "old" quantum mechanics consisting of Einstein's and Planck's theories of quantized transference of electromagnetic energy combined with Bohr's 1913 model of the atom.

Whitehead occasionally refers to the quantum theory in presenting his metaphysical scheme, and it is clear that some, if not most, of these references refer to the "old" quantum theory:

The treatment of cosmology in the philosophy of organism . . . contains the discussion of the ultimate elements from which a more complete philosophical discussion of the physical world—that is to say, of nature—must be derived. In the first place an endeavour has been made to do justice alike to the aspect of the world emphasized by Descartes and to the atomism of the modern quantum theory. Descartes saw the natural world as an extensive spatial plenum, enduring through time. Modern physicists see energy transferred in discrete quanta.[1]

But Whitehead also refers to concepts inherent in the quantum theory as developed by Heisenberg, Bohr, Schrödinger, et al.—the "new" quantum theory, which is formulated, for example, in precisely the same terms Whitehead uses in the quotation from *Process and Reality* that begins this chapter: Namely, "the theory of probability, the theory of tensors, the theory of matrices." (Recall that the combined Hilbert spaces representing a composite system-apparatus-environment are tensor product spaces; and that sets of probability-valuated potentia are grouped into matrices, etc.) His references to this terminology aside, Whitehead also refers to the "new" quantum theory in terms of two fundamental conceptual innovations primarily associated with it: First, the refutation of fundamental materialism as given in Whitehead's "fallacy of misplaced concreteness":

This fallacy consists in neglecting the degree of abstraction involved when an actual entity is considered merely so far as it exemplifies certain categories of thought.[2]

Material substance is one such category, through which the fallacy of misplaced concreteness (and the related "fallacy of undifferentiated endurance") has led to the doctrine of materialism in which

the notion of continuous stuff with permanent attributes, enduring without differentiation and retaining its self-identity though any stretch of time however small or large, has been fundamental. The stuff undergoes change in respect to accidental qualities and relations; but it is numerically self-identical in its character of one actual entity throughout its accidental adventures. The admission of this fundamental metaphysical concept has wrecked the various systems of pluralistic realism. This metaphysical concept has formed the basis of scientific materialism. . . . But this materialistic concept has proved to be as mistaken for the atom as it was for the stone.[3]

Whitehead here refers to the quantum mechanical description of the atom—a description which many physicists, in their debates concerning the proper formulation of the "new" quantum theory, attempted to fit into a materialistic framework. Of these attempts Heisenberg, echoing Whitehead's words above, writes: "It would, in their view, be desirable to return to the reality concept of classical physics or, to use a more general philosophic term, to the ontology of materialism. They would prefer to come back to the idea of an objective real world whose smallest parts exist objectively in the same sense as stones or trees exist." According to the Copenhagen Interpretation, Heisenberg continues, "modern atomic theory no longer allows any reinterpretation or elaboration to make it fit into the naive materialistic conception of the universe."[4] Thus, as Whitehead points out, "the field is now open for the introduction of some new doctrine . . . which may take the place of the materialism with which, since the seventeenth century, science has saddled philosophy."[5]

> What has vanished from the field of ultimate scientific conceptions is the notion of vacuous material existence with passive endurance, with primary individual attributes, and with accidental adventures. Some features of the physical world can be expressed in that way. But the concept is useless as an ultimate notion in science, and in cosmology.[6]

> The simple notion of an enduring substance sustaining persistent qualities, either essentially or accidentally, expresses an abstraction useful for many purposes of life. But whenever we try to use it as a fundamental statement of the nature of things, it proves itself mistaken.[7]

The second conceptual innovation of the "new" quantum theory to which Whitehead refers is closely related to the first: the concept of concrescent state evolution, wherein the final state of a system of facts evolves from the interrelations of its potential facts with the antecedent facts described by the initial system state. This is an exemplification of Whitehead's Ontological Principle:

> Every condition to which the process of becoming conforms in any particular instance, has its reason either in the character of some actual entity in the actual world of that concrescence, or in the character of the subject which is in process of concrescence.[8]

> The actual world is the "objective content" of each new creation.[9]

Whitehead thus characterizes these interrelations between concrescing potentia and the world of actualities antecedent to them as having, in his words, a "vector character"[10] in the sense that each potential fact in the process of concrescence "has its reason" in some particular antecedent fact or facts.[11] Further, the serial evolution of these concrescences—the historical route of state evolutions—manifests itself in the "ultimate vibratory characters of organisms and to the potential element in nature":[12]

> The atom is only explicable as a society with activities involving rhythms with their definite periods. Again the concept shifted its application: protons and electrons were conceived as materialistic electric charges whose activities could be construed as locomotive adventures. . . . The quanta of energy are associated by a simple law with the periodic rhythms which we detect in the molecules. Thus the quanta are, themselves, in their own nature, somehow vibratory; but they emanate from the protons and electrons. Thus there is every reason to believe that rhythmic periods cannot be dissociated from the protonic and electronic entities.[13]

Similarly, in *Modes of Thought* Whitehead writes, "There is a rhythm of process whereby creation produces natural pulsation, each pulsation forming a natural unit of historic fact."[14] The natural unit of historic fact, as applied to quantum mechanics, is the newly evolved, fully determinate system state—a "society" of facts actualized from among a matrix of potential states that have themselves evolved from a society of antecedent facts; and the rhythm is the alternation between the newly evolved, unitary, actual system state and the multiplicity of antecedent facts (and their associated potentia) from which the novel state evolves. Thus, the "many" antecedent facts and their associated potentia become "one" novel state (a novel fact), and are increased, historically, by one—a process which repeats itself "to the crack of doom in the creative advance from creature to creature."[15]

Whitehead's references to the quantum theory as an exemplification of his cosmological scheme, then, pertain to three related concepts—the first one best associated with the "old" quantum theory of Planck and Einstein as applied to Bohr's 1913 atomic model, and the other two best associated with the "new" quantum theory of Heisenberg, Schrödinger, Bohr, et al., typically referred to as the Copenhagen Interpretation:

1. ("old" quantum theory): The transference of electromagnetic energy in discrete quanta and the "vector" relationship between such transference and photonic emissions (Einstein's photoelectric effect). "The mysterious quanta of energy have made their appearance, derived, as it would seem, from the recesses of protons or of electrons. Still worse for the concept, these quanta seem to dissolve into the vibrations of light. Also the material of the stars seems to be wasting itself in the production of the vibrations."[16]

2. ("new" quantum theory): The fallacy of misplaced concreteness (and undifferentiated endurance) as it applies to the doctrine of materialism—successful since the seventeenth century, but unable to accommodate the quantum theory.[17] The latter instead characterizes material substance in terms of "systems" of rhythmically evolving actualities, such that "The atom is only explicable as a society with activities involving rhythms and their definite periods." Thus, "we diverge from Descartes by holding that what he has described as primary *attributes* of physical bodies, are really the forms of internal relationships *between* actual occasions, and *within* actual occasions. Such a change of thought is the shift from materialism to organism, as the basic idea of physical science."[18] This concept is closely related to the next.

3. ("new" quantum theory): The "vector relationship" between potential facts (or systems/societies of facts) in the process of actualization and all antecedent facts, such that the latter contribute in a specific way to the definiteness of the former, in exemplification of the ontological principle: Every actualization of a potential fact is partially determined by its specific relations with all facts antecedent to it, constituting the entire extant universe relative to the actualization at hand. In this way, "the actual world is the 'objective content' of each new creation."[19] The universe is thus characterized as a closed system, as required by any ontologically significant interpretation of the Schrödinger equation, such that any actualization within it necessarily involves all other actualities. This is reflected in Whitehead's "Principle of Relativity," according to which *every actuality* is a potential determinant in the becoming of every new actuality—a principle closely related to his Ontological Principle. The latter is echoed by Heisenberg when he writes, "the transition from the 'possible' to the 'actual' takes place as soon as the interaction of the object with the measuring device, *and thereby with the rest of the*

world, has come into play" (emphasis added).[20] Similarly, the epistemological implications of this fundamental characterization of nature given in the "new" quantum mechanics—the Heisenberg uncertainty relations and Bohr's principle of complementarity—are echoed by Whitehead when, in his commentary on the quantum theory cited above, he writes: "We are now approaching the limits of any reasonable certainty in our scientific knowledge."[21]

Whitehead's references to the quantum theory in *Process and Reality* always reflect what appears to be a distinction between the implications of the "old" quantum theory and those of the "new" quantum theory. Consider the following passages, where concepts best pertaining to the "old" quantum theory are printed in italics, and those best pertaining to the "new" quantum theory are printed in boldface:

> In the language of physical science, the change from materialism to "organic realism"—as the new outlook may be termed—is the displacement of the notion of static stuff by the notion of fluent energy. Such energy has its structure of action and flow, and is inconceivable apart from such structure. **It is also conditioned by "quantum" requirements. These are the reflections into physical science of the individual prehensions, and of the individual actual entities to which these prehensions belong. Mathematical physics translates the saying of Heraclitus, "All things flow," into its own language. It then becomes, All things are vectors.** *Mathematical physics also accepts the atomistic doctrine of Democritus. It translates it into the phrase, All flow of energy obeys "quantum" conditions.*[22]

And:

> [If we] remember that in physics "vector" means definite transmission from elsewhere, we see that this metaphysical description of the simplest elements in the constitution of actual entities agrees absolutely with the general principles according to which the notions of modern physics are framed. **The "datum" in metaphysics is the basis of the vector-theory in physics; the quantitative satisfaction in metaphysics is the basis of the scalar localization of energy in physics.**[23]

And:

> In the language of science, [the philosophy of organism] describes **how the quantitative intensity of localized energy bears in itself the vector marks of its origin, and the specialities of its specific forms;** *it*

also gives a reason for the atomic quanta to be discerned in the building up of a quantity of energy. In this way, the philosophy of organism—as it should—appeals to the facts.[24]

It should be emphasized here that the "new" quantum theory is, in one sense, a more complete systematization of the "old" quantum theory such that the old theory is wholly subsumed within the new theory; therefore, the preceding passages, though they might best apply to either the "old" or the "new" quantum theory, fundamentally apply to both. In the last passage, for example, the "vector marks of [a localized energy's] origin, and the specialities of its specific forms" could also apply to the photoelectric effect of the "old" quantum theory, where energy in the specific form of an emitted photon has a vector relationship to the electron-nucleus interrelation originating the emission.

Discussions of the applicability of Whitehead's philosophy to quantum mechanics typically address whether or not the "new" quantum theory and the "Copenhagen Interpretation" of this theory, formulated contemporaneously with Whitehead's development of his cosmological scheme, might have perhaps influenced the latter in some way. Abner Shimony, for example, states that "Whitehead never refers to the new quantum theory, and it would be unreasonable to expect that even so imaginative a philosopher and scientist as he could have anticipated it except in the most general terms."[25] Shimony proceeds to argue that "the discrepancies . . . between Whiteheadian physics and current microphysics constitute strong disconfirmation of Whitehead's philosophy as a whole."[26] And Henry Folse argues, conversely, that "the philosophy of organism provides a natural context for the acceptance of the Copenhagen Interpretation of quantum theory, especially with the ideas of Bohr and Heisenberg."[27] Folse goes on to say:

> Quite naturally there are many aspects of the philosophy of organism which find no counterpart in the philosophical extrapolations of the Copenhagen Interpretation. . . . There is no reference to the equivalents of "feeling," "satisfaction," or "conceptual prehension." Yet Whitehead would have anticipated this, for the physicists' interpretation of theory is based on a very small segment of experience; Whitehead's system aims at far greater compass. . . .
>
> The Copenhagen position has come under considerable criticism in recent years, much of which draws its strength upon an appeal to

the classical ontology of mechanistic materialism. It would seem that the Copenhagen Interpretation and process philosophy would make good allies in any battle against resurgent substantival materialism. However, the fate of any potential alliance is in jeopardy so long as current discussions of the subject insist on concentrating on the fine points of quantum interpretation rather than its broader more general ramifications.[28]

These points are well taken; however, it has been the purpose of this introductory section to demonstrate that Whitehead did indeed anticipate that the quantum theory would be an exemplification of his cosmological scheme, and not merely vaguely compatible with it. Indeed, to counter Folse's first point above, Whitehead goes so far as to suggest specific correlations between the nomenclature of his scheme and that of quantum theory:

> If we substitute the term "energy" for the concept of a quantitative emotional intensity, and the term "form of energy" for the concept of "specific form of feeling," and remember that in physics "vector" means definite transmission from elsewhere, *we see that this metaphysical description of the simplest elements in the constitution of actual entities agrees absolutely with the general principles according to which the notions of modern physics are framed.* The "datum" in metaphysics is the basis of the vector-theory in physics; the quantitative satisfaction in metaphysics is the basis of the scalar localization of energy in physics; the "sensa" in metaphysics are the basis of the diversity of specific forms under which energy clothes itself. . . . *The general principles of physics are exactly what we should expect as a specific exemplification of the metaphysics required by the philosophy of organism.*[29] (emphasis added)

Whitehead's claim here should not be overlooked; the metaphysical scheme he presents in *Process and Reality* and in other writings was absolutely intended to be a fundamental characterization of nature as exemplified by the theoretical physics of his time, which included the development of modern quantum mechanics. Aside from his explicitly saying so, the overwhelming detail in which he presents his cosmological model is more than indicative of that intention. For how could it be that the fundamental features of his philosophy are reflected and analyzable in practically every aspect of human experience, as thoroughly elaborated upon in his writings, *except* that of the physics of his time? If any aspect of human experience closely correlates with the specific features of Whitehead's metaphysical scheme, it should certainly be the latter—especially given that the quantum theory is a purely mathematical theory, and therefore clearly within the technical scope of Whitehead's expertise.

It is the purpose of this chapter to demonstrate how quantum mechanics, as given by the modern decoherence-based interpretations described thus far[30] is, in the most specific terms of the mechanics, an extremely precise, phase-by-phase exemplification of Whitehead's cosmological model. It is an exemplification both conceptually and mechanically, and in terms of physical nature, quantum mechanics is thus the most fundamental exemplification of Whiteheadian metaphysics currently capable of analysis.

The most general correlation between quantum mechanics and the Whiteheadian cosmological system pertains to the concept of state evolution in the former, and concrescence in the latter. These terms describe the same process as elaborated in Part I, wherein:

(i) A world of existing, mutually interrelated facts (Whitehead's "actual occasions") is presupposed.

(ii) The inclusion of these facts (Whitehead's "positive prehension" or "feeling" of facts as "data") in the act of measurement or state specification of them—by their necessary mutual interrelations, somehow entails:

    (*a*) All other facts and their associated potentia—either in their inclusion in the specification, or their necessary exclusion from specification. This requirement is reflected in Whitehead's "Principle of Relativity" and his "Ontological Principle," and in quantum mechanics, by the Schrödinger equation's exclusive applicability to closed systems, with the universe being the only such system.[31] The exclusions relate to the process of negative selection productive of the decoherence effect described earlier, and Whitehead refers to these eliminations as "negative prehensions." Their form and function with respect to environmental degrees of freedom are, as we shall see, identical to those related to the process of decoherence.

    (*b*) The evolution of the system of all facts into a novel fact—namely, a maximal specification (the "state" specification) of the relevant facts (those not excluded by decoherence or "negatively prehended" in Whitehead's terminology). State specification—the maximal specification of many facts via

the necessary exclusion of some facts—thus entails the evolutionary production of a novel fact—namely, a unification
of the facts specified.

(c) The requirement that this evolution proceed relative to a particular fact, typically belonging to a particular subsystem of
facts. In quantum mechanics, these are referred to, respectively, as the "indexical eventuality" and the "measuring apparatus"; Whitehead's equivalent term is, simply, the
prehending "subject." This requirement is given in Whitehead's "Ontological Principle" and "Category of Subjective
Unity"; their correlates in quantum mechanics—the necessary relation of a state evolution to some "preferred basis"
characteristic of the measuring apparatus—has often been
misapprehended as a principle of sheer subjectivity, the
source of the familiar lamentations that quantum mechanics
destroys the objective reality of the world.

(iii) Measurement or state specification thus entails, at its heart, the
anticipated actualization (or "concrescence") of one novel potential fact/entity from many valuated potential facts/entities which
themselves arise from antecedent facts (data); and it is understood that the quantum mechanical description of this evolution
terminates in a matrix of probability valuations, anticipative of
a final unitary reduction to a single actuality. Ultimately, then,
concrescence/state evolution is unitary evolution from actualities
to unique actuality. But when analyzed into subphases, both concrescence and state evolution entail a fundamental nonunitary evolution, analogous to von Neumann's conception of quantum
mechanics as most fundamentally a nonunitary state reduction
productive of a unitary reduction.[32] It is an evolution from:

(a) a multiplicity of the actual many—that is, $|\Psi\rangle$, to

(b) a matrix of potential "formal" (in the sense of applying a
"form" to the facts) integrations or unifications of the many
(Whitehead's term is propositional "transmutations" of the
many—a specialized kind of "subjective form"—and he also
groups these into "matrices"[33]). Each of these potential integrations is described in quantum mechanics as a projection
of a vector representing the actual, evolving multiplicity of
facts onto a vector (or subspace) representing a potential
"formally integrated" eigenstate. The Whiteheadian analog
of the actual multiplicity's "projection" onto a potential integration is "ingression"—where a potential formal integration

arises from the ingression of a specific "potentiality of definiteness"[34] via a "conceptual prehension" of that specific potentiality of definiteness (Whitehead also refers to these potentia as "eternal objects," and explicitly equates the two terms[35]). But whereas in quantum mechanics, the state vector representing the actual multiplicity of facts is *projected onto* the potential integration (the eigenvector representing the eigenstate), in Whitehead's scheme it is the latter which *ingresses into* the prehensions of the actual multiplicity. This difference reflects Whitehead's concern with the origin of these potentia; according to his Ontological Principle, if they ingress into the evolution, they must be thought of as coming from somewhere. The eigenstate, or object of projection in quantum mechanics, is, in contrast, simply extant. This difference aside, Whiteheadian vector "ingression" and quantum mechanical vector "projection" are conceptually equivalent—as are the terms "eternal object" and "potential fact."

There are, furthermore, two important characteristics shared by both the quantum mechanical and Whiteheadian notions of potentia. First, there is a sense in which both are "pure" potentia, referent to no specific actualities. For Whitehead, "eternal objects are the pure potentials of the universe; and the actual entities differ from each other in their realization of potentials."[36] "An eternal object is always a potentiality for actual entities; but in itself, as conceptually felt, it is neutral as to the fact of its physical ingression in any particular actual entity of the temporal world."[37] In quantum mechanics, this pure potentiality is reflected in the fact that the state vector $|\Psi\rangle$ can be expressed as the sum of an infinite number of vectors belonging to an infinite number of subspaces in an infinite number dimensions, representing an infinite number of potential states or "potentialities of definiteness," referent to no specific actualities and potentially referent to all. Many of these are incapable of integration, forming nonsensical, interfering superpositions, and are eliminated as negative prehensions in a subsequent phase of concrescence.

Second, quantum mechanical projections are also "inherited" from the facts constituting the initial state of the system (as well as the historical route of all antecedent states subsumed by the initial state) such that preferred bases in quantum mechanics are typically reproduced in the evolution from state to state. Similarly, in Whitehead's scheme, antecedent facts, when prehended, are typically "objectified" by one of their own historical "potential forms of definiteness"—usually the potential forms that were antecedently actualized at some point in the historical route of occasions constituting the system measured.

> An actual entity arises from decisions *for* it and by its very existence provides decisions *for* other actual entities which supersede it.[38]

> Some conformation is necessary as a basis of vector transition, whereby the past is synthesized with the present. The one eternal object in its two-way function, as a determinant of the datum and as a determinant of the subjective form, is thus relational. . . . An eternal object when it has ingression through its function of objectifying the actual world, so as to present the datum for prehension, is functioning "datively."[39]

Whitehead's characterization of potentia as "relational" is exemplified by the manner in which potentia mediate the actuality of a measured system and the actuality of the outcome of the measurement—that is, the mediation of the initial and final system states.

The evolution thus continues into its next phase:

> (c) A reintegration of these integrations into a matrix of "qualified propositional" transmutations,[40] involving a process of negative selection where "negative prehensions" of potentia incapable of further integration are eliminated. The potential unifications or propositional transmutations in this reduced matrix are each qualified by various valuations. Each potential transmutation relative to the indexical eventuality of the measuring apparatus (i.e., each potential outcome state relative to the apparatus and some prehending subject belonging to it) is thus a potential "form" into which the potential facts will ultimately evolve. Whitehead terms these "subjective forms" and as applied to quantum mechanics, the term "subjective" refers to the fact that the "form" of each potential outcome state is reflected in the preferred basis relative to the indexical eventuality of the measuring apparatus (i.e., the prehending subject). Again, it is only the form that is thus subjective—for any number of different devices with different preferred bases could be used to measure a given system, and any number of different people with their own "mental preferred bases" could interpret (measure) the different read-

ings of the different devices, and so on down the von Neumann chain of actualizations. The potential facts to which each subjective form pertains, however, are initially "given" by the objective facts constitutive of the world antecedent to the concrescence at hand. Thus, again, the "subjective form" of a preferred basis is in no way demonstrative of sheer subjectivity—that is, the evolution of novel facts *as determined* by a particular subject; it is, rather, demonstrative of the evolution of novel facts *jointly* determined by both the world of facts antecedent to the evolution *and* the character of the subject prehending this evolution by virtue of its inclusion in it. Again, according to Whitehead's "Ontological Principle,"

> Every condition to which the process of becoming conforms in any particular instance, has its reason either in the character of some actual entity in the actual world of that concrescence, or in the character of the subject which is in process of concrescence.[41]

> The actual world is the "objective content" of each new creation.[42]

The evolution thus proceeds to and terminates with what Whitehead terms the "satisfaction" which in quantum mechanical terms is described as:

(*d*) The actualization of the final outcome state—that is, one subjective form from the reduced matrix of many subjective forms—in "satisfaction" of the probability valuations of the potential outcome states in the reduced matrix. In quantum mechanics, as in Whitehead's model, this actualization is irrelevant and transparent apart from its function as a datum (fact) in a subsequent measurement, such that the "prehending subject" becomes "prehended superject." Again, this is simply because both Whitehead's process of concrescence and quantum mechanics presuppose the existence of facts, and thus cannot account for them. For Whitehead, "satisfaction" entails "the notion of the 'entity as concrete' abstracted from the 'process of concrescence'; it is the outcome separated from the process, thereby losing the actuality of the atomic entity, which is both process and outcome."[43] Thus, the probability valuations of quantum mechanics describe probabilities that a given potential outcome state *will be actual* upon observation—thus implying a subsequent evolution and an interminable evolution of such evolutions. Every fact or system of facts in quantum mechanics, then, subsumes and implies both an initial state and a final state; there can be no state specification S without reference, implicit or explicit, to $S_{initial}$ and S_{final}. This is reflected in Whitehead's scheme by referring to the "subject" as the "subject-superject":

> The "satisfaction" is the "superject" rather than the "substance" or the "subject." It closes up the entity; and yet is the superject adding its character to the creativity whereby there is a becoming of entities superseding the one in question. . . .[44]

> An actual entity is to be conceived as both a subject presiding over its own immediacy of becoming, and a superject which is the atomic creature exercising its function of objective immortality. . . . [45]

> It is a subject-superject, and neither half of this description can for a moment be lost sight of. . . .[46]

> [The superject is that which] adds a determinate condition to the settlement for the future beyond itself.[47]

Thus, the process of concrescence is never terminated by actualization/satisfaction; it is, rather, both begun and concluded with it. The many facts and their associated potentia become one novel state (a novel fact), and are thus increased, historically, by one, so that "the oneness of the universe, and the oneness of each element in the universe, repeat themselves to the crack of doom in the creative advance from creature to creature."[48] "The atomic actualities individually express the genetic unity of the universe. The world expands through recurrent unifications of itself, each, by the addition of itself, automatically recreating the multiplicity anew."[49]

The Phases of Quantum Mechanical Concrescence

> The process of concrescence is divisible into an initial stage of many feelings, and a succession of subsequent phases of more complex feelings integrating the earlier simpler feelings, up to the satisfaction which is one complex unity of feeling. This is the "genetic" analysis of the

> satisfaction. The actual entity is seen as a process; there is growth from phase to phase; there are processes of integration and reintegration.
>
> Alfred North Whitehead, *Process and Reality*

The process of concrescence described here by Whitehead is exemplified by the process of quantum mechanical state evolution, which entails the evolution of a closed system of objectively extant facts (data) correlative to the concrescent evolution of a single subject-fact (the indexical eventuality of the measuring apparatus). This evolution consists of a succession of phases in which manifold potential specifications of each datum are integrated—relative to the measuring apparatus and, particularly, its preferred basis. This integration entails the elimination of incompatible and irrelevant specifications via a process of negative selection, productive of the decoherence effect. The result is a matrix of decoherent, mutually exclusive and exhaustive alternative, probability-valuated integrations or potential system states. Of these, one will become actual fact in satisfaction of these valuations as revealed retrodictively by future measurements in its role as datum for these measurements.

> The final phase in the process of concrescence . . . is termed the "satisfaction." It is fully determinate (a) as to its genesis, (b) as to its objective character for the transcendent creativity [i.e., as datum for subsequent actualizations], and (c) as to its prehension, positive or negative, of every item in its universe. . . .[50]
>
> The satisfaction is merely the culmination marking the evaporation of all indetermination.[51]

Furthermore, the evolution from phase to phase in a concrescence does not occur in asymmetrically modal "physical time," and this is exemplified in quantum mechanics by the time-independent Schrödinger equation which most fundamentally describes this evolution. Each phase, rather, presupposes the actualization as a quantum whole—further exemplified in quantum mechanics, for example, by the inability to specify "what happens between" one observation and the next. This principle, an infamously mystifying component of quantum mechanics for many, is entirely intuitive when interpreted according to the Whiteheadian scheme: For apart from facts, there is nothing to specify.

> This genetic passage from phase to phase is not in physical time: the exactly converse point of view expresses the relationship of concres-

cence to physical time. The actual entity is the enjoyment of a certain quantum of physical time. But the genetic process is not the temporal succession: such a view is exactly what is denied by the epochal theory of time. Each phase in the genetic process presupposes the entire quantum, and so does each feeling in each phase. . . . It can be put shortly by saying, that physical time expresses some features of the growth, but *not* the growth of the features.[52]

Whitehead divides the process of concrescence into three "stages";[53] the first two stages, Stage I, the responsive stage or "physical pole," and Stage II, the supplementary stage, or "mental pole," entail the integrations of prehensions (or "concrete facts of relatedness") of antecedent facts, and it is these two stages that find their exemplification in quantum mechanical state reduction. Stage III is termed the "satisfaction," which is the actualization of one of the many potential integrations generated in the first two stages. Stages I and II together consist of three successive phases (see figure 4.1, back end sheet), which are precisely analogous to the three phases of state evolution (the valuation of potentia) in orthodox "Copenhagen" quantum mechanics: (i) initial state, which evolves to become (ii) an integration of potentia subsumed by the pure state density matrix, which evolves to become (iii) a reintegration of potentia subsumed by the mixed state, reduced density matrix. Phases (*b*) and (*c*) here, as given in the Copenhagen formalism, entail two concepts also present in Whitehead's Phase 2 and Phase 3 in the supplementary stage, as will be explored presently: (i) the transitional nonunitary state evolution suggested by von Neumann; (ii) the process of negative selection productive of the decoherence effect related to this nonunitary state reduction.

It must be emphasized that these two additional concepts in no way require the modification of the orthodox quantum formalism with any additional features (i.e., linear or nonlinear modifications of the Schrödinger equation as suggested by Wigner, for example, or more recently, as suggested in the Spontaneous Localization interpretation of Ghirardi, Rimini, and Weber). All that is required is the orthodox quantum formalism, but applied universally—in satisfaction of the requirement of ontological coherence and consistency— rather than as conventionally and instrumentally applied only to specific situations incapable of accommodation by classical mechanics or ontology.

Whitehead continues:

There are three successive phases of feelings, namely, a phase of "conformal" feelings [Phase 1, which belongs to Stage I], one of "conceptual" feelings [Phase 2, which belongs to Stage II], and one of "comparative" feelings [Phase 3, which belongs to Stage II], including "propositional" feelings in this last species. . . . The two latter phases can be put together as the "supplemental" [stage].[54]

Phases 1, 2, and 3 and their relationship to Stages I, II, and III, have been diagramed on the back end sheet of this book in order to facilitate the following discussion. The first phase in Whitehead's scheme—the only phase in Stage I, the "responsive stage" or alternately, the "physical pole"—is termed the "primary phase." It is a "conformal," "responsive" phase, wherein the actual world as a multiplicity of facts is initially related to ("prehended" by) the subject. These facts constitute the whole of the antecedent universe relative to the concrescence—that is, facts comprised by the initial state of a measured system, together with that of its measuring apparatus and environment in accord with the closed-system requirement of the Schrödinger equation. This requirement in the ontological interpretation of quantum mechanics is an exemplification of Whitehead's "Principle of Relativity":

The potentiality for being an element in a real concrescence of many entities into one actuality, is the one general metaphysical character attaching to all entities, actual and non-actual [i.e., actual and potential]. Every item in its universe is involved in each concrescence. In other words, it belongs to the nature of a "being" that it is a potential for every "becoming." This is the "principle of relativity."[55]

The principle of universal relativity directly traverses Aristotle's dictum, "(A substance) is not present in a subject." On the contrary, according to this principle an actual entity *is* present in other actual entities. In fact if we allow for degrees of relevance, and for negligible relevance, we must say that every actual entity is present in every other actual entity. The philosophy of organism is mainly devoted to the task of making clear the notion of "being present in another entity." This phrase is here borrowed from Aristotle: it is not a fortunate phrase, and in subsequent discussion it will be replaced by the term "objectification."[56]

According to Whitehead's cosmological scheme as exemplified by the ontological interpretation of quantum mechanics, then, in every concrescence (or "actualization of a potential fact," or occasion of "state specification" in a measurement interaction) every fact constitutive of the universe is an initial datum which typically becomes "objectified" by one of its own historical potentia. Often, as in most cases of physical transmission of energy, the objectified potential is the one that was actualized in its own process of concrescence. (Recall from Part I that it is this regular reproduction of antecedent potentia throughout historical routes of facts that contributes to the "classicality" of high-inertia systems.) Objectified data are "positively prehended" in Whitehead's terminology, or equivalently, "felt." Thus, the primary phase is the seat of causal efficacy, via "simple physical feelings," in the process of concrescence.

> A simple feeling has the dual character of being the cause's feeling reenacted for the effect as subject. But this transference of feeling effects a partial identification of cause with effect, and not a mere representation of the cause. It is the cumulation of the universe and not a stage play about it. By reason of this duplicity in a simple feeling there is a vector character which transfers the cause into the effect. . . . Simple physical feelings embody the reproductive character of nature, and also the objective immortality of the past.[57]

Quantum mechanics echoes this conception of the causal interrelations between past facts and the becoming of novel facts. Recall the quote from Omnès:

> Past facts are not absolutely real; they only *were* real. One can never indicate a past fact by pointing a finger at it and saying "that." One must call for a memory or use a record, a note, a photograph. Nevertheless, the derivation of a unique past is possible because quantum mechanics allows for the existence of memory and records. . . . What we observe in reality is always something existing *right now*, even if we interpret it as a trace of an event in the past, whether it be a crater on the moon, the composition of a star atmosphere, or the compared amounts of uranium and lead in a rock.[58]

And in the same spirit, Charles Hartshorne comments, "It is quantum theory that has at last brought science to admit the contingency that qualifies every instance of becoming."[59]

In summary of the primary phase of concrescence, Whitehead writes:

> The primary stage in the concrescence of an actual entity is the way in which the antecedent universe enters into the constitution of the

entity in question, so as to constitute the basis of its nascent individuality. . . .[60]

The first phase is the phase of pure reception of the actual world in its guise of objective datum for aesthetic synthesis. . . .[61]

The actual world is the "objective content" of each new creation.[62]

The second and third phases of concrescence in Whitehead's scheme together constitute the "supplementary stage" or alternatively, the "mental pole,"[63] and whereas the primary stage/physical pole is the seat of causal efficacy, the supplementary stage/mental pole is the seat of causal indeterminism, via "conceptual prehensions" or prehensions of potentia, qualified as two types, "real potentiality" and "general potentiality" (also referred to as "pure potentiality"):

> Thus we have always to consider two meanings of potentiality: (a) the "general" potentiality, which is the bundle of possibilities, mutually consistent or alternative, provided by the multiplicity of eternal objects, and (b) the "real" potentiality, which is conditioned by the data provided by the actual world. General potentiality is absolute, and real potentiality is relative to some actual entity, taken as a standpoint whereby the actual world is defined.[64]

A "real potentiality" is potentiality conditioned by the extant facts of the universe to which they pertain—that is, potential facts historically embodied and thus conditioned by the actual universe from which they have evolved. The potential outcome of a measurement of the color status of our red-green traffic signal from Part I, for example, is conditioned by the inheritance of antecedent system states embodied by the history of the traffic-signal system, correlated with the histories of the measuring apparatus and the environment englobing both—that is, the history describing the universe itself. The two "real potentia" of this particular measurement are, then, "red light" and "green light" (assuming these are the only two potential colors to ever have been actualized throughout the history of that traffic signal). It is via such "real potentia" that the causal prehensions from the primary phase are objectified as "immanent," historically *realized* determinants of the facts prehended. But in the supplementary stage of the currently evolving concrescence, these potentia are now also transcendent capacities for determination; for if our traffic signal were determined via measurement to be red just prior to the current measurement, the outcome might nevertheless be green this time.

Each entity in the universe of a given concrescence *can*, so far as its own nature is concerned, be implicated in that concrescence in one or other of many modes; but *in fact* it is implicated only in *one* mode. . . . The particular mode of implication is only rendered fully determinate by that concrescence, though it is conditioned by the correlate universe. This indetermination, rendered determinate in the real concrescence, is the meaning of "potentiality." It is a *conditioned* indetermination, and is therefore called a *"real* potentiality."[65]

Real potentia so defined reflect Whitehead's "Category of Conceptual Reproduction"—one of his nine "Categoreal Obligations," which are necessarily presupposed by the process of concrescence, as well as by the ontological interpretation of quantum mechanics described here. The Category of Conceptual Reproduction is described by Whitehead thus:

From each physical feeling there is the derivation of a purely conceptual feeling whose datum is the eternal object exemplified in the definiteness of the actual entity, or the nexus, physically felt.[66]

Conceptual Reproduction (or equivalently, "Conceptual Valuation") constitutes the first subphase of Phase 2—namely, the beginning of the supplementary stage/mental pole. And whereas the conditioned indetermination of "real potentia" in this first subphase provides conditional transcendent novelty to an evolving fact, the unconditioned indetermination of "pure potentia" provides unconditioned transcendent novelty. A "pure potentiality" (or "eternal object")—a "pure potential for the specific determination of fact"[67]—"does not involve a necessary reference to any definite actual entities of the temporal world. . . . An eternal object is always a potentiality for actual entities; but in itself, as conceptually felt, it is neutral as to the fact of its physical ingression in any particular actual entity of the temporal world."[68]

An eternal object can be described only in terms of its potentiality for "ingression" into the becoming of actual entities; and that its analysis only discloses other eternal objects. It is a pure potential. The term "ingression" refers to that particular mode in which the potentiality of an eternal object is realized in a particular actual entity, contributing to the definiteness of that actual entity.[69]

The unconditioned indetermination provided by pure potentia, like the conditioned indetermination of real potentia, reflects another Categoreal Obligation: the "Category of Conceptual Reversion" which occurs in the second subphase of Phase 2:

> There is a secondary origination of conceptual feelings with data which are partially identical with, and partially diverse from, the eternal objects forming the data in the first phase of the mental pole. . . .
>
> Thus the first phase of the mental pole is conceptual reproduction, and the second phase is a phase of conceptual reversion. . . . It is the category by which novelty enters the world; so that even amid stability there is never undifferentiated endurance. But, as the category states, reversion is always limited by the necessary inclusion of elements identical with elements in feelings of the antecedent phase.[70]

> Note that [the Category of Conceptual Reproduction] concerns conceptual reproduction of physical feeling, and [the Category of Conceptual Reversion] concerns conceptual diversity from physical feeling.[71]

In Whitehead's metaphysical scheme and in quantum mechanics, alternative potential integrations of data (alternative specifications of facts belonging to the system measured)—that is, alternative "forms" of specified factual relations (prehensions of data)—are always relative to a single evolving fact (the prehending subject/indexical eventuality of the measuring apparatus); and thus, these alternative forms of integration—to use Whitehead's term—are "subjective forms." The subjective forms of both real and pure potentia are "valuations"—that is, they have a qualified intensity in a given concrescence—"valuation up" ("adversion") or "valuation down" ("aversion").[72] Thus, the subjective form of a "real" potential—that is, a Conceptual Reproduction—is either valuated "up" or "down" compared to other alternative subjective forms of alternative real potentia, which are similarly valuated. Subjective forms of Conceptual Reversions are also valuated "up" or "down."

In quantum mechanics, these valuations are exemplified as probability valuations by which each potential outcome state (i.e., each alternative potential integration of specified facts belonging to a measured system together with its apparatus) is qualified. Thus, the subjective form "traffic signal = green" is a potential integration of facts belonging to the traffic signal, unified according to a "real" potential (light = green) previously actualized in the historical traffic-signal system. "Traffic signal = green" is, then, a Conceptual Reproduction. In quantum mechanics, this subjective form (potential outcome state) is valuated "up" or "down" according to a probability between zero and one. And similarly, the subjective form "traffic signal = blue" is a potential integration of facts belonging to the traffic signal, unified according to a "pure" potential not previously actualized in the historical traffic signal system. "Traffic signal = blue," then, is a Conceptual Reversion. In quantum mechanics, even the subjective form of a conceptual reversion is valuated "up" or "down" according to a probability between zero and one. Recall the calculation Murray Gell-Mann described making as an undergraduate wherein he predicted the probability that a heavy macroscopic object would leap a foot into the air of its own accord; the subjective form of that particular Conceptual Reversion was valuated as a probability of $1 / 1 \times 10^{62}$—most definitely an "aversion" or "valuation down" in relation to the subjective forms of the Conceptual Reproductions pertaining to that calculation, one of which was "the object remains at rest."

This distinction between pure and real potentia, including their respective roles in the first two subphases of the supplementary stage—Conceptual Reproduction and Conceptual Reversion—have precise analogs in quantum mechanics; in fact, the representation of potentia as vectors in Hilbert space provides an excellent model in aid of understanding the functional relationship between real and pure potentia in Whitehead's metaphysical scheme. Recall that our idealized red-green traffic signal system was described via a Hilbert space of two dimensions, represented by x–y coordinate axes, such that a vector of unit length along each axis represents the state of the system S as S_{green} or S_{red}, respectively. S_{green} and S_{red} are thus "real potentia" of the system. One may even generalize that it is a system's "real potentia," correlated with those of the indexical eventuality of the measuring apparatus, which determine the number of mutually orthogonal vectors—and therefore the number of dimensions—required by a Hilbert space representing the system. Recall that mutually orthogonal vectors represent mutually exclusive potential states—a requirement of Whiteheadian potentia: "The definiteness of the actual arises from the exclusiveness of eternal objects in their function as determinants. If the actual entity be *this*, then by the nature of the case it is not *that* or *that*. The fact of incompatible alternatives is the ultimate fact in virtue of which there is definite character."[73]

These two sentences together convey the very essence of quantum mechanics. We have already seen their mathematical exemplification: If the mutually orthogonal vectors $|u_{green}\rangle$ and $|u_{red}\rangle$ represent "real" mutually exclusive potential states of the traffic signal system, then $|\psi\rangle$—a vector of unit length which belongs to neither the subspace $\mathscr{E}_{green}$ nor $\mathscr{E}_{red}$—represents the potential ingression of all pure potentia. For $|\psi\rangle$ can be expressed as the sum of a practically infinite number of vectors, each representing a purely potential state, referent to no particular facts or history of facts; thus some of these purely potential states are sensible and capable of integration in the current evolution, and some are not. Those which are *must* be mutually exclusive according to both quantum mechanics and Whitehead's cosmological scheme. These are the "real potentia"—in this case, $|u_{green}\rangle$ and $|u_{red}\rangle$, which in terms of the pure potentiality inherent in $|\psi\rangle$, are expressed:

$$|\psi\rangle = \alpha|u_{green}\rangle + \beta|u_{red}\rangle$$

where $|\alpha|^2 + |\beta|^2 = 1$.

One of these real potentia $|u_{green}\rangle$ or $|u_{red}\rangle$ will become actual, according to its valuation as expressed by the complex coefficient α or β. Both $|u_{green}\rangle$ and $|u_{red}\rangle$, as real potentia, are (i) realized determinants of facts historically actualized, and in that sense, are "immanent"; and (ii) capacities for determination in the current evolution, and in this sense, "transcendent."[74]

The representation of pure potentia (Whitehead's "eternal objects") in the expression $|\psi\rangle = \alpha|u_{green}\rangle + \beta|u_{red}\rangle$ is disclosed when it is recalled that α and β—the valuations of the real potentials $|u_{green}\rangle$ and $|u_{red}\rangle$ respectively—are *projections* of $|\psi\rangle$ onto $|u_{green}\rangle$ and $|u_{red}\rangle$ respectively. In Whiteheadian terms, this is the *ingression* of both the real potentia $|u_{green}\rangle$ and $|u_{red}\rangle$ into the concrescence, along with the other pure potential states embodied by $|\psi\rangle$ (again, represented mathematically by its potential expression as the sum of a practically infinite number of vectors). Recall that the state vector $|\psi\rangle$, referred to as a "pure state" in quantum mechanics, represents a pure potential referent at once to no specific facts *and* to all facts. Yet it is an objectively *real* state, for it is also a maximal specification of the bare multiplicity of facts constituting the system in its initial state—that is, facts determined antecedent to the concrescence at hand. Thus, the expression $|\psi\rangle = \alpha|u_{green}\rangle + \beta|u_{red}\rangle$ represents the dipolar nature of the concrescence given in both the primary stage of conditioned causal determination, and in the supplementary stage of conditioned indetermination.

> The process of becoming is dipolar, (i) by reason of its qualification by the determinateness of the actual world, and (ii) by its conceptual prehensions of the indeterminateness of eternal objects. The process is constituted by the influx of eternal objects into a novel determinateness of feeling which absorbs the actual world into a novel actuality.[75]

Clearly, however, $|\psi\rangle$, given its component of pure potentiality, can be expressed as the sum of a multiplicity of vectors and projections other than those contained in the integration $\alpha|u_{green}\rangle + \beta|u_{red}\rangle$. These coherent superpositions of potentia are eliminated, and it is this elimination which, in a subsequent phase, will produce the two decoherent, mutually exclusive and exhaustive real potential outcome states represented by $\alpha|u_{green}\rangle + \beta|u_{red}\rangle$. The Whiteheadian account of this elimination and its mechanism will be discussed presently, but for now can be inferred from the two sentences cited above: "The fact of incompatible alternatives is the ultimate fact in virtue of which there is definite character." Or, in the words of Hartshorne, "An actual world cannot be all possible worlds. . . . To be actual is to exclude some possibilities."[76]

> The initial problem is to discover the principles according to which some eternal objects are prehended positively and others are prehended negatively. Some are felt and others are eliminated.[77]

Before exploring the concept and mechanism of these eliminations in the form of negative prehensions, we must provide a more accurate mathematical expression for the process of concrescence thus far; for $|\psi\rangle = \alpha|u_{green}\rangle + \beta|u_{red}\rangle$ only represents the integrations of prehensions. The concrescing subject-superject (i.e., the indexical eventuality belonging to a measuring apparatus) is only implied. All that is required is an expression which correlates the concrescent evolution of the measured (prehended) system with that of the measuring apparatus, relative to the indexical eventuality/prehending subject. The following expression, introduced and discussed at length in Part I, does this:

$$|\Phi^c\rangle = \alpha|u_{green}\rangle|d_\downarrow\rangle + \beta|u_{red}\rangle|d_\uparrow\rangle$$

But as we have seen, the elimination of superpositions of incompatible potentia in quantum mechanics requires the correlation of facts belonging to the evolving system-apparatus with those belonging to the whole of the environment which englobes them. These environmental correlations are also required by any coherent ontological interpretation of quantum mechanics, which applies only to closed systems such as that of a universe characterized as system-apparatus-environment. Whitehead holds the same requirement in his metaphysical scheme, where the concrescent evolution of any one potential fact or system ("nexus") of such facts necessarily entails all facts antecedent to the evolution. And as is the case in quantum mechanics, Whitehead's elimination via negative prehensions of incompatible potentia—the "thwarting elements of a nexus"[78]— occurs by virtue of the elimination of the "unwelcome detail"[79] generated by these necessary environmental correlations—precisely the same mechanism described by the decoherence-based interpretations of quantum mechanics discussed in Part I. We can now proceed to explore in more detail these conceptual similarities between Whitehead's process of negative selection and that of the decoherence effect in quantum mechanics, as the environmental correlations required by both are accounted for in the expression:

$$|\Phi^c\rangle|E_0\rangle = (\alpha|u_{green}\rangle|d_\downarrow\rangle + \beta|u_{red}\rangle|d_\uparrow\rangle)|E_0\rangle$$
$$\Rightarrow \alpha|u_{green}\rangle|d_\downarrow\rangle|E_\downarrow\rangle + \beta|u_{red}\rangle|d_\uparrow\rangle|E_\uparrow\rangle = |\Psi\rangle$$

where $|\Psi\rangle$ is the system (nexus) of all facts prehended as two "real," and therefore mutually exclusive,[80] potential integrations $\alpha|u_{green}\rangle|d_\downarrow\rangle|E_\downarrow\rangle$ and $\beta|u_{red}\rangle|d_\uparrow\rangle|E_\uparrow\rangle$, each one a valuated "subjective form" of $|\Psi\rangle$— that is, an integration relative to the prehending subject/indexical eventuality $|d\rangle$ belonging to the detector. Thus, we have a mathematical expression of the process of concrescence (excluding the "satisfaction"),

$$|\Psi\rangle = \alpha|u_{green}\rangle|d_\downarrow\rangle|E_\downarrow\rangle + \beta|u_{red}\rangle|d_\uparrow\rangle|E_\uparrow\rangle$$

which describes

a growth of prehensions, with integrations, eliminations, and determination of subjective forms. But the determination of successive phases of subjective forms, whereby the integrations have the characters that they do have, depends on the unity of the subject imposing a mutual sensitivity upon the prehensions.[81]

This equation is conceptually similar to that describing pure state discussed above—

$$|\psi\rangle = \alpha|u_{green}\rangle + \beta|u_{red}\rangle$$

—only now correlated with facts belonging to both the detector and environment, in satisfaction of the requirements of ontological coherence.

THE CATEGOREAL OBLIGATIONS

Five Whiteheadian "Categoreal Obligations" are operative thus far in our examination of the quantum mechanical exemplification of concrescence—two of which we have already discussed (the "Category of Conceptual Reproduction" and the "Category of Conceptual Reversion")—and it will be useful to describe the other three here prior to discussing the process of negative selection, which will introduce several others. It is important to note that these Categoreal Obligations are not only exemplified in quantum mechanics; they are presupposed by quantum mechanics. The first three pertain to the primary stage of concrescence.

Categoreal Obligation I: The Category of Subjective Unity

The many feelings which belong to an incomplete phase in the process of an actual entity, though unintegrated by reason of the incompleteness of the phase, are compatible for synthesis by reason of the unity of their subject.[82]

Quantum mechanics describes the evolution of the state of a system as a set of alternative integrations of potential facts born of antecedent "actual" facts, and these integrations are always relative to a particular fact—the indexical eventuality—as it also evolves. It is this state evolution relative to a single evolving eventuality and its preferred basis which allows these integrations to occur; and it is by virtue of this relativity that all potentia involved in the evolution will be compatible, and that those which are not will be eliminated.

The novel actual entity, which is the effect, is the reproduction of the many actual entities of the past. But in this reproduction there is abstraction from their various totalities of feeling. This abstraction is required by the categoreal conditions for compatible synthesis in the novel unity. The limitation, whereby the actual entities felt are severally reduced to the perspective of one of their own feelings, is imposed by the categoreal condition of subjective unity, requiring a harmonious compatibility in the feelings of each incomplete phase. This subjective insistence may, from the beginning, replace positive feelings by negative prehensions. Feelings are dismissed by negative prehensions, owing to their lack of compliance with categoreal demands.[83]

Categoreal Obligation II: The Category of Objective Identity

There can be no duplication of any element in the objective datum of an actual entity, so far as concerns the function of that element in the satisfaction. . . . This category asserts the essential self-identity of any entity as regards its status in each individualization of the universe. In such a concrescence one thing has one rôle, and cannot assume any duplicity. This is the very meaning of self-identity, that, in any actual confrontation of thing with thing, one thing cannot confront itself in alien rôles. Any one thing remains obstinately itself playing a part with self-consistent unity. This category is one ground of incompatibility.[84]

The elimination of superpositions of interfering potentia in quantum mechanics is an exemplification of this category, which obliges the satisfaction of the logical principle of noncontradiction. Simply, it is by the Category of Objective Identity that a confused creature like Schrödinger's cat cannot exist objectively in its superposed state.

Categoreal Obligation III: The Category of Objective Diversity

There can be no "coalescence" of diverse elements in the objective datum of an actual entity, so far as concerns the functions of those elements in that satisfaction. . . . "Coalescence" here means the notion of diverse elements exercising an absolute identity of function, devoid of the contrasts inherent in their diversities.[85]

The third category is concerned with the antithesis to oneness, namely, diversity. An actual entity is not merely one; it is also definitely complex. But, to be definitely complex is to include definite diverse elements in definite ways. The category of objective diversity expresses the inexorable condition—that a complex unity must provide for each of its components a real diversity of status, with a reality which bears the same sense as its own reality and is peculiar to itself. In other words, a real unity cannot provide sham diversities of status for its diverse components.[86]

Category III is particularly important to any ontological interpretation of quantum mechanics which characterizes potentia as a species of reality rather than merely an artifact of some epistemic boundary that cannot be crossed. For in an interpretation which maintains the latter, the quantum mechanical evolution of a system is fundamentally a system of subatomic, material locomotions. What evolves is merely our knowledge of these locomotions, and therefore the diverse potentia "integrated" in the evolution are "sham diversities of status"—vacuous, statistical, necessarily approximate specifications of these locomotions, where the "status" of one such potential is equivalent to the status of any.

In the interpretation of quantum mechanics described here, however, the ontological significance of potentia presupposes Category III; for if potentia are indeed a species of reality, then each potential must be as unique as the specific actuality from which it evolves. And therefore integrations of potentia are "real diversities."

Because Category IV, the Category of Conceptual Reproduction, and Category V, the Category of Conceptual Reversion were already introduced above, we can now move on to Category VI.

Categoreal Obligation VI: The Category of Transmutation

Returning to our idealized traffic signal system, the quantum mechanical expression of concrescence is given as:

$$|\Psi\rangle = \alpha|u_{green}\rangle|d_\downarrow\rangle|E_\downarrow\rangle + \beta|u_{red}\rangle|d_\uparrow\rangle|E_\uparrow\rangle$$

This expression, in both the decoherence-based interpretations of quantum mechanics and in Whitehead's scheme, presupposes a process of negative selection via negative prehensions. This process, as we saw in Part I, is expressed via the use of matrices of potentia—or more precisely, matrices of alternative integrations of potentia. These integrations are exemplifications of the Whiteheadian "subjective forms" of potentia, given that these potentia are expressed in terms of $|d\rangle$—that is, relative to $|d\rangle$ as the prehending subject/indexical eventuality belonging to the detector. The pure state density matrix contains all the ingressed potentia, pure and real, related to the data of the correlated system-apparatus-environment.

$$\rho^c = |\Psi\rangle\langle\Psi|$$
$$= |\alpha|^2 |u_{green}\rangle\langle u_{green}||d_\downarrow\rangle\langle d_\downarrow||E_\downarrow\rangle\langle E_\downarrow|$$
$$+ \boldsymbol{\alpha\beta} |u_{green}\rangle\langle u_{red}||d_\downarrow\rangle\langle d_\uparrow||E_\downarrow\rangle\langle E_\uparrow|$$
$$+ \boldsymbol{\alpha^*\beta} |u_{red}\rangle\langle u_{green}||d_\uparrow\rangle\langle d_\downarrow||E_\uparrow\rangle\langle E_\downarrow|$$
$$+ |\beta|^2 |u_{red}\rangle\langle u_{red}||d_\uparrow\rangle\langle d_\uparrow||E_\uparrow\rangle\langle E_\uparrow|$$

Some of these potentia are capable of integration into the mutually exclusive subjective forms $\alpha|u_{green}\rangle|d_\downarrow\rangle|E_\downarrow\rangle$ and $\beta|u_{red}\rangle|d_\uparrow\rangle|E_\uparrow\rangle$; and others are incapable of such integration because of their mutual nonexclusivity (interference) and thus constitute coherent superpositions of potentia. These are represented by the two middle terms in boldface and are the nonsensical subjective forms that are eliminated from the concrescing evolution by virtue of the excess prehensions of environmental data they include—again, "excess" relative to the prehending subject/indexical eventuality. These excess potentia are eliminated in the Whiteheadian scheme via

a massive average objectification of a nexus, while eliminating the detailed diversities of the various members of the nexus in question. This method, in fact, employs the device of blocking out unwelcome detail. It depends on the fundamental truth that objectification is abstraction. It utilizes this abstraction inherent in objectification so as to dismiss the thwarting elements of a nexus into negative prehensions. At the same time the complex intensity in the structured society is supported by the massive objectifications of the many environmental nexūs, each in its unity as one nexus, and not in its multiplicity as many actual occasions.

This mode of solution requires the intervention of mentality operating in accordance with the Category of Transmutation. It ignores diversity of detail by overwhelming the nexus by means of some congenial uniformity which pervades it. The environment may then change indefinitely so far as concerns the ignored details—so long as they can be ignored.

The close association of all physical bodies, organic and inorganic alike . . . suggests that this development of mentality [i.e., activity of the "Mental Pole" or "Supplementary Phase"—not conscious mentality] is characteristic of the actual occasions which make up the structured societies we know as "material bodies."[87]

The mathematical analog of the Category of Transmutation is, as we have seen, the "trace-over" or "sum-over" of the potentia in the pure state density matrix such that the "detailed diversities" stemming from the environmental correlations, represented by the off-diagonal terms, are eliminated:

$$\rho_{SD} \equiv \text{Tr}_E |\Psi\rangle\langle\Psi| = \Sigma_i \langle E_i |\Psi\rangle\langle\Psi| E_i\rangle = \varrho^r$$

This yields a reduced density matrix ϱ^r of mutually exclusive and exhaustive subjective forms, or valuated "real" potential integrations of the system-apparatus nexus

$$\rho^r = |\alpha|^2 |u_{green}\rangle\langle u_{green}| |d_\downarrow\rangle\langle d_\downarrow| + |\beta|^2 |u_{red}\rangle\langle u_{red}| |d_\uparrow\rangle\langle d_\uparrow|$$

and evinces the "Principle of Relativity" so crucial to Whitehead's cosmology: Every fact is a potential determinant in the becoming of every new fact. This reduced density matrix is merely another way of expressing the earlier equation:

$$|\Phi^c\rangle = \alpha |u_{green}\rangle |d_\downarrow\rangle + \beta |u_{red}\rangle |d_\uparrow\rangle$$

This "massive average objectification" is possible mathematically for the same reason it is possible conceptually in Whitehead's scheme; it is prefaced by an integration of a vast multiplicity of potential outcomes (owing to the manifold degrees of freedom contributed by the environment) into various "subjective forms" relative to the prehending subject/indexical eventuality $|d\rangle$. In the case of our traffic light example, this yields two subjective forms:

(i) $|\alpha|^2 |u_{green}\rangle\langle u_{green}| |d_\downarrow\rangle\langle d_\downarrow| |E_\downarrow\rangle\langle E_\downarrow|$

(ii) $|\beta|^2 |u_{red}\rangle\langle u_{red}| |d_\uparrow\rangle\langle d_\uparrow| |E_\uparrow\rangle\langle E_\uparrow|$

Each subjective form is a potential integration of all facts belonging to the nexus of facts comprising the system, the detector, and the environment relative to the prehending subject/indexical eventuality $|d\rangle$ belonging to the detector. Each potential integration, by the manifold environmental correlations, thus entails a practically infinite number of potentia—some prehended positively, others negatively—each one related to a separate environmental datum. Each of these two subjective forms consists of manifold potential integrations of these prehensions of the system-apparatus-environment nexus, but each shares a common, "defining characteristic" determined by each subjective form—namely, the particular status of the indexical eventuality $|d\rangle$. In mathematical terminology, each subjective form is thus referred to as an "equivalence class" or "coarse-grained" integration. It is this grouping into equivalence classes or Whiteheadian "subjective forms" that allows for the cancellations among ignored "detailed diversities," thus "blocking out unwelcome detail" (or "coarse graining" these details); and it is by these cancellations that the elimination of the interfering, incompatible potentia, represented by the two terms in boldface on p. 149, is effected.

> The irrelevant multiplicity of detail is eliminated, and emphasis is laid on the elements of systematic order in the actual world. [88]

> In this process, the negative prehensions which effect the elimination are not merely negligible. The process through which a feeling passes in constituting itself, also records itself in the subjective form of the integral feeling. The negative prehensions have their own subjective forms which they contribute to the process. A feeling bears on itself the scars of its birth; it recollects as a subjective emotion its struggle for existence; it retains the impress of what it might have been, but is not. It is for this reason that what an actual entity has avoided as a datum for feeling may yet be an important part of its equipment. The actual cannot be reduced to mere matter of fact in divorce from the potential. [89]

> The right co-ordination of negative prehensions is one secret of mental progress; but unless some systematic scheme of relatedness characterizes the environment, there will be nothing left whereby to constitute vivid prehension of the world. [90]

Again, this "right co-ordination of negative prehensions" via a "systematic scheme of relatedness" characterizing the environment, productive of correlate coordinations of positive prehensions, occurs in Phase 3 of the concrescence in accord with the Category of Transmutation, which "ignores diversity of detail by overwhelming the nexus by means of some congenial uniformity which pervades it. The environment may then change indefinitely so far as concerns the ignored details—so long as they can be ignored." [91] The process of decoherence, as described above—via integrations of potentia into "equivalence classes" or "subjective forms" according to a defining

characteristic—is a precise exemplification of the Category of Transmutation; for it describes the integration of a bare multiplicity of
facts into a potential system (nexus) where each potential fact in
the potential system shares a "defining characteristic" given in the
indexical eventuality of the measuring apparatus. Thus, each potential integration of this kind (a nexus with "social order") is exemplified mathematically as an "equivalence class" or "coarse-grained"
integration:

> A "society" . . . is a nexus with social order. A nexus enjoys "social
> order" where (i) there is a common element of form illustrated in the
> definiteness of each of its included actual entities, and (ii) this com-
> mon element of form arises in each member of the nexus by reason of
> the conditions imposed upon it by its prehensions of some other
> members of the nexus, and (iii) these prehensions impose that condi-
> tion of reproduction by reason of their inclusion of positive feelings of
> that common form. Such a nexus is called a "society," and the com-
> mon form is the "defining characteristic" of the society. The notion
> of "defining characteristic" is allied to the Aristotelian notion "sub-
> stantial form."[92]

And as this concept applies to the Category of Transmutation, defined by Whitehead:

> When (in accordance with category IV, or with categories IV and V)
> one, and the same, conceptual feeling is derived impartially by a pre-
> hending subject from its analogous simple physical feelings of various
> actual entities, then in a subsequent phase [Phase 3] of integra-
> tion—of these simple physical feelings together with the derivate con-
> ceptual feeling—the prehending subject may transmute the *datum* of
> this conceptual feeling into a characteristic of some *nexus* containing
> those prehended actual entities, or of some part of that nexus; so that
> the nexus (or its part), thus qualified, is the objective datum of a
> feeling entertained by this prehending subject. Such a transmutation
> of simple physical feelings of many actualities into one physical feeling
> of a nexus as one, is called a "transmuted feeling."[93]

> It is evident that the complete datum of the transmuted feeling is a
> contrast, namely, "the nexus, as one, in contrast with the eternal ob-
> ject." This type of contrast is one of the meanings of the notion "qual-
> ification of physical substance by quality."[94]

The decoherence effect in quantum mechanics—via coarse-
grained, alternative "equivalence classes" productive of the necessary eliminations of interfering potentia—is an exemplification of

the Category of Transmutation both conceptually and "mechanically": Both the decoherence effect and the Category of Transmutation play an identical role in accounting for the logically ordered, enduring characteristics of the "classical" macroscopic world, including our notion of "qualification of physical substance by quality"; and each ultimately provides the conceptual and mechanical means by which its fundamental characterization of nature can apply to human experience.

The distinction between "conceptual" and "mechanical" means is important here, for Whitehead had the former in mind with respect to the Category of Transmutation more so than the latter; he was not aware, for example, of the recent quantum mechanical interpretations which use the decoherence effect to *mechanically* account for the classicality of macroscopic objects. But even though Whitehead was more interested in the Category of Transmutation conceptually as it applies to perception (i.e., its use in accounting for perceptive errors[95]), he explicitly refrained from predicating Transmutation upon consciousness:

> It is evident that adversion and aversion, and also the Category of Transmutation, only have importance in the case of high-grade organisms. They constitute the first step towards intellectual mentality, though in themselves do not amount to consciousness.[96]

Nevertheless, the application of the Category of Transmutation to even the most rudimentary actual occasions—as is suggested here in its relation to the quantum mechanical decoherence effect—would likely not be objected to by Whitehead, for he explicitly proposes the function of Transmutation—and Reversion—in quantum mechanics. Indeed, it is this fundamental physical role of transmutation from which higher-order mental transmutations ultimately derive:

> The physical theory of the structural flow of energy has to do with the transmission of simple physical feelings from individual actuality to individual actuality. Thus some sort of quantum theory in physics, relevant to the existing type of cosmic order, is to be expected. The physical theory of alternative forms of energy, and of the transformation from one form to another form, ultimately depends upon transmission conditioned by some exemplification of the Categories of Transmutation and Reversion.[97]

> Apart from transmutation our feeble intellectual operations would fail to penetrate into the dominant characteristics of things. We can only

understand by discarding. . . . Transmutation is the way in which the actual world is felt as a community, and is so felt in virtue of its prevalent order. For it arises by reason of the analogies between the various members of the prehended nexus, and eliminates their differences.[98]

Transmutation in Phase 3 of the concrescence, as exemplified in quantum mechanics by the decoherence effect, results in a reduced density matrix of mutually exclusive, "coarse-grained" subjective forms, each valuated as a probability. Thus, in the case of our idealized traffic signal system, each of the terms in the reduced density matrix

(i) $|\alpha|^2 |u_{green}\rangle\langle u_{green}||d_\downarrow\rangle\langle d_\downarrow|$

(ii) $|\beta|^2 |u_{red}\rangle\langle u_{red}||d_\uparrow\rangle\langle d_\uparrow|$

represents a subjective form or real potential integration of the nexus of facts comprising the system and detector. Excess environmental correlations have been eliminated, carrying with them the coherent superpositions of incompatible potentia incapable of integration. Again, each subjective form is a potential integration (potential societal nexus) of the traffic signal system/detector facts relative to a particular indexical eventuality/prehending subject, specified by $|d\rangle$.

Transmutation entails the "generic contrast" of (i) the real potential serving as the "defining characteristic" by which a given social nexus ("equivalence class") is integrated (in our example, the defining characteristic would be the indexical eventuality $|d_\uparrow\rangle$ or $|d_\downarrow\rangle$); and (ii) the reality of the system of facts ($|\Psi\rangle$) being prehended (related) in the concrescence (measurement). In quantum mechanics, this is exemplified by the contrast of (i) one of the alternative "real potential," probable measurement outcome states in the reduced matrix (again, "state" being a maximal specification or form of the facts belonging to a system); and (ii) the system of actual, extant facts as data. It is a contrast, in other words, of potential forms with actual facts. This contrast is, once again, mathematically represented as

$$|\Psi\rangle = \alpha |u_{green}\rangle |d_\downarrow\rangle |E_\downarrow\rangle + \beta |u_{red}\rangle |d_\uparrow\rangle |E_\uparrow\rangle$$

and, in Whitehead's scheme, is productive of "comparative feelings," of which "physical purposes" are one type:

> The integration of each simple physical feeling [in Phase 1] with its conceptual counterpart [in Phase 2] produces in a subsequent phase

[Phase 3] a physical feeling whose subjective form of re-enaction has gained or lost subjective intensity according to the valuation up, or the valuation down, in the conceptual feeling. This is the phase of physical purpose.[99]

In the integral comparative feeling the datum is the contrast of the conceptual datum with the reality of the objectified nexus. The physical feeling [or "Concrete Fact of Relatedness"] is feeling a real fact; the conceptual feeling is valuing an abstract possibility. . . . This synthesis of a pure abstraction $[\alpha|u_{green}\rangle|d_\downarrow\rangle|E_\downarrow\rangle$ or $\beta|u_{red}\rangle|d_\uparrow\rangle|E_\uparrow\rangle$] with a real fact $[|\Psi\rangle]$, as in feeling, is a generic contrast. . . .

The constancy of physical purposes explains the persistence of the order of nature. . . . [In a] physical purpose, the datum is the generic contrast between the nexus, felt in the physical feeling, and the eternal object valued in the conceptual feeling. This eternal object is also exemplified as the pattern of the nexus. Thus the conceptual valuation now closes in upon the feeling of the nexus as it stands in the generic contrast, exemplifying the valued eternal object. This valuation accorded to the physical feeling endows the transcendent creativity with the character of adversion, or of aversion. The character of adversion secures the reproduction of the physical feeling, as one element in the objectification of the subject beyond itself. Such reproduction may be thwarted by incompatible objectification derived from other feelings. But a physical feeling, whose valuation produces adversion, is thereby an element with some force of persistence into the future beyond its own subject. It is felt and re-enacted down a route of occasions forming an enduring object. . . .

When there is aversion, instead of adversion, the transcendent creativity assumes the character that it inhibits, or attenuates, the objectification of that subject in the guise of that feeling. Thus aversion tends to eliminate one possibility by which the subject may itself be objectified in the future. Thus adversions promote stability; and aversions promote change without any indication of the sort of change.[100]

There are two species of physical purpose: (i) physical purposes due to conceptual reproduction; and (ii) physical purposes due to conceptual reversion. A physical purpose of the first species, such as those facts related to the transfer of energy, merely "receives the physical feelings, confirms their valuations according to the 'order' of that epoch, and transmits by reason of its own objective immortality. Its own flash of autonomous individual experience is negligible."[101] And it is because of "the origination of reversions in the [supplementary stage] . . . that vibration and rhythm have a dominating importance in the physical world"[102]—a concept clearly exemplified by objective quantum indeterminacy.

Any ontological interpretation of quantum mechanics wherein potentia are characterized as a species of reality must at some point address not only the relationship between these potentia and the actualities from which they evolve, but also the relationship between these potentia and the actualities they evolve to become. Phases 1 and 2, and their correlate Categoreal Obligations, primarily concern the former; Phase 3—the phase of comparative feelings and transmutation—is primarily concerned with the latter. It concerns not only what these potential and probable outcome states are in quantum mechanics, and from whence they evolved—questions related to Phases 1 and 2; but also *why* these potential and probable outcome states are what they are. And as Whitehead's metaphysical scheme entails answers to other ontological questions commonly raised with respect to quantum mechanics, such as those explored thus far, it entails an answer to this important question as well.

The answer is to be found in the three final Categoreal Obligations, which pertain to the balance of adversion and aversion—reproduction and reversion—regularity and diversity in the process of concrescence. Whitehead suggests that a tendency toward such balance of adversion and aversion permeates all of nature and is qualitatively manifest as "balanced complexity"—and perhaps quantitatively as well, if indeed quantum mechanics is an exemplification of Whitehead's cosmological scheme.

Categoreal Obligation VII: The Category of Subjective Harmony

The valuations of conceptual feelings are mutually determined by the adaptation of those feelings to be contrasted elements congruent with the subjective aim. Category I and category VII jointly express a pre-established harmony in the process of concrescence of any one subject. Category I has to do with the data felt, and category VII with the subjective forms of the conceptual feelings. This pre-established harmony is an outcome of the fact that no prehension [or "Concrete Fact of Relatedness"] can be considered in abstraction from its subject, although it originates in the process creative of its subject.[103]

By the Category of Subjective Unity (category I), and by the seventh category of Subjective Harmony, all origination of feelings is governed by the subjective imposition of aptitude for final synthesis. In the former category the intrinsic inconsistencies, termed "logical" [i.e., superpositions of nonexclusive, interfering potentia], are the formative conditions in the pre-established harmony. In this seventh category, and in the Category of Reversion, aesthetic adaptation for the end is the formative condition in the pre-established harmony. . . . It is only by reason of the Categories of Subjective Unity, and of Subjective Harmony, that the process constitutes the character of the product, and that conversely the analysis of the product discloses the process.[104]

Categoreal Obligation VIII: The Category of Subjective Intensity

The subjective aim, whereby there is origination of conceptual feeling, is at intensity of feeling (α) in the immediate subject, and (β) in the relevant future. This double aim—at the *immediate* present and the *relevant* future—is less divided than appears on the surface. For the determination of the *relevant* future, and the *anticipatory* feeling respecting provision for its grade of intensity, are elements affecting the immediate complex of feeling. . . . The relevant future consists of those elements in the anticipated future which are felt with effective intensity by the present subject by reason of the real potentiality for them to be derived from itself.[105]

Categoreal Obligation IX: The Category of Freedom and Determination

The concrescence of each individual actual entity is internally determined and is externally free. . . . This category can be condensed into the formula, that in each concrescence whatever is determinable is determined, but that there is a remainder for the decision of the subject-superject of that concrescence. *The subject-superject is the universe in that synthesis, and beyond it there is nonentity.* This final decision is the reaction of the unity of the whole to its own internal determination. . . . But the decision of the whole arises out of the determination of the parts, so as to be strictly relevant to it.[106]

It is thus by the Category of Subjective Harmony that aversions productive of diversities are the seeds of complexity rather than chaos. And it is by the Category of Subjective Intensity that a "subjective aim" toward complexity—the balance between regularity and diversity, reproduction and reversion—is, in part, generated by an anticipatory feeling of the relevant future. It might seem at first glance that quantum mechanics, and indeed any aspect of physics, ought solely concern itself with the immediate present as causally related to the past; but the fact that quantum mechanics terminates with probability valuations clearly evinces the relevance of Categoreal Obligations VII and VIII. For the notion of *relevant future* is a defining element in the concept of mathematical probability. Unlike a classical measurement, which is solely a qualification of the immediate present (which, of course, may be used to *predict* the relevant future, but does not in itself necessarily entail the relevant future), a quantum mechanical measurement yields probabilities, which necessarily pertain to both the immediate present *and* the relevant future.

As discussed earlier, the concept of a probability-valued outcome state in quantum mechanics necessarily presupposes both facts antecedent to the measurement, as well as facts subsequent to and consequent of the measurement. That probabilities are potentia qualified as mutually exclusive and exhaustive is related to the Category of Subjective Unity; that probabilities are potentia qualified as valuations—tendencies toward reproduction or reversion—is related to the Category of Subjective Harmony and Subjective Intensity.

> It follows that balanced complexity is the outcome of this final category of subjective aim. Here "complexity" means the realization of contrasts, of contrasts of contrasts, and so on; and "balance" means the absence of attenuations due to the elimination of contrasts which some elements in the pattern would introduce and other elements inhibit. . . .[107]

> It requires a complex constitution to state diversities as consistent contrasts. . . . The contrasts produced by reversion are contrasts required for the fulfillment of the aesthetic ideal. Unless there is complexity, ideal diversities lead to physical impossibilities, and thence to impoverishment. . . .[108]

> Thus there is urge towards the realization of the maximum number of eternal objects [pure potentia] subject to the restraint that they must be under conditions of contrast.[109]

The study of complexity in nature has been the topic of a great deal of research and debate over the past several years, and it is very likely that quantum mechanics describes such complexity at the most fundamental physical level. The concept of "balanced complexity" suggested by Whitehead, and its function and exemplification in his metaphysical scheme, has a direct analog in the concept of "effective complexity"—also a balance of regularity and diversity. The function and exemplification of effective complexity in quantum mechanics will likely prove to be an important avenue of inquiry in the coming years, and the decoherence-based interpretations of quantum mechanics are clearly the interpretations best suited to such an investigation. For the decoherence effect is predicated upon the very notions of contrast elucidated in Categories VII and VIII—*diverse* multiplicities of facts, contrasted with *regulated* potential integrations of these facts—notions that clearly pertain to "balanced complexity" in nature. One might, for example, correlate the "urge towards the realization of the maximum number of eternal objects subject to the restraint that they must be under conditions of contrast" with an urge toward expressing $|\Psi\rangle$ in terms of the maximum number of potential eigenvectors, subject to the restraint of necessary contrast, such that they be mutually orthogonal and thus representative of mutually exclusive, exhaustive, potential outcome states.

NOTES

1. Alfred North Whitehead, *Process and Reality: An Essay in Cosmology, Corrected Edition*, ed. D. Griffin and D. Sherburne (New York: Free Press, 1978), 238–239.

2. Ibid., 7.

3. Ibid., 78.

4. Werner Heisenberg, *Physics and Philosophy* (New York: Harper Torchbooks, 1958), 129.

5. Alfred North Whitehead, *Science and the Modern World* (New York: Free Press, 1967), 47.

6. Whitehead, *Process and Reality*, 309.

7. Ibid., 79.

8. Ibid., 24.

9. Ibid., 65.

10. Ibid., 237.

11. Ibid., 116.

12. Ibid., 239.

13. Ibid., 78–79.

14. Alfred North Whitehead, *Modes of Thought* (New York: Macmillan, 1938), 120.

15. Whitehead, *Process and Reality*, 228.

16. Ibid., 78–79.

17. This could also apply, though less obviously, to the "old" quantum theories of Einstein and Planck.

18. Whitehead, *Process and Reality*, 309.

19. Ibid., 65.

20. Heisenberg, *Physics and Philosophy*, 55–56.

21. Whitehead, *Process and Reality*, 78.

22. Ibid., 309.

23. Ibid., 116.

24. Ibid., 117.

25. Abner Shimony, "Quantum Physics and the Philosophy of Whitehead," *Boston Studies in the Philosophy of Science*, ed. R. Cohen and M. Wartofsky (New York: Humanities Press, 1965), 2:308.

26. Ibid., 322.

27. Henry Folse, "The Copenhagen Interpretation of Quantum Theory and Whitehead's Philosophy of Organism," *Tulane Studies in Philosophy* 23 (1974): 33.

28. Ibid., 47.

29. Whitehead, *Process and Reality*, 116.

30. That is, with the explicit inclusion of von Neumann's concept of nonunitary state reduction (his "Process 1") and the related mechanism of decoherence, which requires nothing more than the usual Schrödinger equation of the Copenhagen formalism.

31. That is, ontologically or "truly" closed; one can easily create "epistemologically" closed systems, as is often done in the laboratory, where a given experimental arrangement can—without the need of any complicated physical procedure—be effectively "closed" from distant and effectively unrelated environmental facts. The purpose of the ontological interpretation of quantum mechanics is to disclose the mechanism by which such epistemologically closed systems are created, and why such excluded factual relations are necessary. The decoherence effect is one such mechanism being investigated by physicists.

32. John von Neumann, *Mathematical Foundations of Quantum Mechanics* (Princeton, N.J.: Princeton University Press, 1955).

33. Whitehead, *Process and Reality*, 285.

34. Ibid., 40.

35. Ibid., 149.

36. Ibid., 149.

37. Ibid., 44.

38. Ibid., 43.

39. Ibid., 164.

40. Ibid., 285–286.

41. Ibid., 24.

42. Ibid., 65.

43. Ibid., 84.

44. Ibid., 84.

45. Ibid., 45.

46. Ibid., 29.

47. Ibid., 150.

48. Ibid., 228.

49. Ibid., 286.

50. Ibid., 25–26.

51. Ibid., 212.

52. Ibid., 283.

53. Ibid., 212.

54. Ibid., 164.

55. Ibid., 22.

56. Ibid., 50.

57. Ibid., 237–238.

58. Roland Omnès, *The Interpretation of Quantum Mechanics* (Princeton, N.J.: Princeton University Press, 1994), 344.

59. Charles Hartshorne, "Bell's Theorem and Stapp's Revised View of Space-Time," *Process Studies* 7, no. 3 (1977): 188.

60. Whitehead, *Process and Reality*, 152.

61. Ibid., 212.

62. Ibid., 65.

63. The term "mental pole" is not intended to suggest that consciousness or any sort of intellectual function be ascribed to all facts; nor should the terms "physical pole" and "mental pole" be correlated with the dualism of Descartes, especially given that the Whiteheadian philosophy is a fundamental repudiation of this dualism.

64. Whitehead, *Process and Reality*, 65.

65. Ibid., 23.

66. Ibid., 26.

67. Ibid., 22.

68. Ibid., 44.
69. Ibid., 23.
70. Ibid., 249.
71. Ibid., 26.
72. Ibid., 253–254.
73. Ibid., 240.
74. Ibid., 239–240.
75. Ibid., 45.
76. Hartshorne, "Bell's Theorem," 188.
77. Whitehead, *Process and Reality*, 248.
78. Ibid., 101.
79. Ibid., 101.
80. The mutual exclusivity of "real" potentia (i.e., probable outcome states) is, again, a requirement of both the Whiteheadian scheme and quantum mechanics. Whitehead writes: "The definiteness of the actual arises from the exclusiveness of eternal objects in their function as determinants. If the actual entity be *this*, then by the nature of the case it is not *that* or *that*. The fact of incompatible alternatives is the ultimate fact in virtue of which there is definite character."
81. Whitehead, *Process and Reality*, 235.
82. Ibid., 26.
83. Ibid., 238.
84. Ibid., 225.
85. Ibid., 225.
86. Ibid., 227.
87. Ibid., 101.
88. Ibid., 254.
89. Ibid., 226–227.
90. Ibid., 254.
91. Ibid., 101.
92. Ibid., 34.
93. Ibid., 251.
94. Ibid., 27.
95. Ibid., 253.
96. Ibid., 254.
97. Ibid., 254.
98. Ibid., 251.
99. Ibid., 248–249.
100. Ibid., 276–277.
101. Ibid., 245.
102. Ibid., 277.
103. Ibid., 27.

104. Ibid., 255.
105. Ibid., 27.
106. Ibid., 27–28.
107. Ibid., 278.
108. Ibid., 255.
109. Ibid., 278.

5

Spatiotemporal Extension

FOR BOTH WHITEHEADIAN METAPHYSICS and the decoherence-based interpretations of quantum mechanics, physical objects—whether they be solid material bodies or localized fields of energy—are most fundamentally characterized as serial historical routes of quantum actual occasions. Nonlocal correlations among physical objects, such as those encountered in the EPR-type experiments discussed earlier, are described simply as logically necessary correlations among the *histories* constituting spatially well separated object systems. The logical necessity of these correlations derives from the understanding that any local system is necessarily participant in a broader environmental system, whose history subsumes those of all the local systems within it. In this way, the broader environmental history imposes consistency conditions, such as those first suggested by Robert Griffiths,[1] upon the manifold local histories it includes as these histories unfold, quantum event by quantum event. These consistency conditions are rooted ultimately in the logical principles of noncontradiction and the excluded middle.

The broadest conceivable system of actualities is the universe itself—the ultimate environment for any actuality and the ultimate history—and its provision of consistency conditions upon the histories of the local systems subsumed within it evinces a universal holism largely incompatible with the mechanistic materialism underlying classical physics. For it is not only a conceptual holism that one might confine to the realms of spiritual, philosophical, or theological tradition; it is also a physically significant holism—at least as regards the decoherence effect and those interpretations of quantum mechanics that employ it.

Because quantum mechanics exemplifies certain holistic features of the universe and divulges the logical historical relations among all actualities, it might be tempting to conclude that *any aspect* of physical relations among actualities must derive from, or reduce to, these holistic logically ordered features. What is emphasized in such a conclusion is the universal scope of any single, brute, quantum fact. When the local history describing a physical system is augmented by a novel quantum fact, the entire system history is changed, as is the history of any system included within it, or any system including it, irrespective of their spatiotemporal metrical topologies. The temptation, in other words, is to regard the spatiotemporal extensiveness of actualities and systems of actualities, as well as any theory describing such extensiveness (such as Einstein's special and general theories of relativity), as mere abstractions derivable entirely from fundamental quantum events, in the same way that the concept of "material body" is so derivable. Various information theoretic approaches to quantum mechanics emphasize these holistic features, some to the extent of redefining the whole of physics as a reduction entirely to relations among quantum events as sheerly logically ordered quantum "information bits."[2]

The logical, historical relation among all actualities and the associated holistic features of the universe are, as we have seen, essential components of Whiteheadian metaphysics—components that form the bulk of its connection to modern quantum theory. It is only natural, then, that information theoretic approaches to quantum mechanics that elevate these holistic features to primacy, and relegate spatiotemporal extension to mere abstraction, might be portrayed as being compatible with Whiteheadian metaphysics. The "fallacy of misplaced concreteness" would thus find its exemplification not only in the conventional notion of "fundamental materiality" but also in the conventional notion of "fundamental extensiveness in spacetime." But for Whitehead, the spatiotemporally *extensive* morphological structure ("coordinate division") of actualities and nexūs of actualities, as well as of spacetime itself, is as crucial to concrescence as the *intensive* features of their relations ("genetic division") emphasized by the information theoretic interpretations of quantum mechanics. The "intensive" features are the logical, historical, genetic relations operative in the supplementary stage of concrescence as described in quantum mechanical terms in previous chapters. "Intensity" is reflected in the valuations of subjective forms, or in quantum mechanical terms, in the probability valuations of potential outcome states. Indeed, there is a direct relationship between (i) the *extensive* morphological structure of actualities objectified as spatiotemporally coordinated data in the primary stage of concrescence (the "physical pole") and (ii) the *intensive* valuations of the subjective forms of these data in the supplementary stage (the "mental pole")—valuations that emphasize the logical, historical, genetic relations of the objectified data.

This close relationship is a key aspect of the dipolarity of concrescence and the avoidance of a Cartesian bifurcated Nature. The world is not merely a multiplicity of disembodied quantum physical facts upon which individual subjects project their own relativistic and vacuous spatiotemporal coordinations. For Whitehead, these extensive coordinations are local nexūs and societies of actual occasions—"relativistic" in the sense that these occasions *are* the relata—related to each other and to the prehending subject by virtue of their spatiotemporally coordinated objectification by this subject and according to the subject's own particular spatiotemporal morphology (i.e., its particular spacetime reference frame). And by the Category of Transmutation in the supplementary stage, these spatiotemporal coordinations in the primary stage contribute to the determination of "environmental" data to be eliminated as negative prehensions—an elimination key to the quantum mechanical decoherence effect discussed previously, and to the valuation of the subjective forms of potential quantum mechanical outcome states. The extensive relations of the physical pole and the intensive relations of the mental pole are, then, closely interwoven, such that the operation of neither pole can be abstracted from the operation of the other. Thus the logical coherence of the dipolarity of concrescence is evinced. The extensive coordination of actualities cannot occur apart from the logically prior historical genesis of the actualities coordinated; and the historical genesis of actualities cannot occur apart from the logically prior extensive coordination of the actualities ingredient in the novel genesis. The only ontology capable of accommodating such dipolarity is one wherein all beings are fundamentally historical and perpetual routes of quantum becomings.

COORDINATE DIVISION AND GENETIC DIVISION

The relativistic extensive relations among the actualities disclosed by coordinate division are relations of causal efficacy; the phenomenon of local *causal influence of actualization by temporally prior actu-*

ality is derived from these extensive relations. And the phenomenon of nonlocal *causal affection of potentia by logically prior actuality* is derived from the logical, historical relations among actualities disclosed by genetic division as correlated with quantum mechanics in previous chapters. Whitehead tends to restrict the term "intensity of relations" to genetic division, in reference to the valuations (i.e., quantum mechanical probability valuations), adversions, and aversions of subjective forms in the supplementary stage of concrescence. This restriction of "intensity of relations" gives contrast to the "extensiveness of relations" pertaining to coordinate division operative in the primary stage. But because concrescence is dipolar, with the operations of both poles being mutually implicative (as opposed to "complementary" or "bipolar," where the operations of both poles would be mutually exclusive), it is clear that the intensive probability valuations in the supplementary stage/mental pole are closely related to the extensive spatiotemporal coordinations of the primary stage/physical pole.

Recall, for example, the story of the traveling salesman in Hong Kong whose wife in California is about to give birth. The question, "At what moment does the salesman become a father?" has only two significant possible answers: (a) at the moment his wife gives birth, for this moment is a common event in each of their histories; (b) as soon as a signal sent from the spacetime coordinates of the birth, traveling no faster than the speed of light, reaches the spacetime coordinates of the father; for prior to that, he cannot be causally altered by the event.

The first answer reflects the genetic analysis of the events—the logical, historical, and thus "nonlocal" relations between the salesman and his wife; it entails an implicit characterization of them as historically correlated quantum mechanical systems. The mutual nonlocality, or any other extensive qualities of these systems, has absolutely no bearing on the *fact* of the birth event augmenting each history in a correlated way; and once those potentia associated with the salesman's *history* are affected—his "history" defining him not only by his past, but also by the potentia associated with his future (what he might be, might do, might be able to do, might be for others, etc.)—then in some sense *he* is affected, whether he is aware of the affection or not. The phenomenon associated with this first answer and its analogous EPR-type nonlocal quantum mechanical correlations is "causal affection of potentia by logically prior actuality." (Again, an information theoretic approach to quantum mechanics would emphasize this feature over any spatiotemporally extensive qualifications of the systems.) But the *intensity* of the correlation, in quantum mechanical/Whiteheadian terms, derives in part from the extent to which (i) the wife-system and its birth event, and (ii) the salesman-system, are entangled *extensively*, that is, *spatiotemporally coordinately*, with a shared environment—another spatiotemporally coordinated system, whose global history subsumes their local histories, as well as the local histories of every other event or system of events within it.

Recall that apart from the logical consistency conditions provided by a global environmental history, crucial to the decoherence process, one would be left with a bare superposition of practically infinite potential outcome histories of negligible individual intensity, belonging to a practically infinite number of spatiotemporally disconnected events. But because of these historical environmental consistency conditions, a vast number of individual potential outcome histories are eliminated by negative selection and a process of mutual cancellation—a physical exemplification of the logical principle of noncontradiction. The superposition of this multiplicity of minimally intensive potential outcome states thus evolves, via the negative selection process of decoherence and transmutation, to become a reduced matrix of probability-valuated, mutually exclusive and exhaustive outcome states or propositional transmutations, each valuation reflective of a significant *intensity*. Thus, in both the decoherence-based interpretations of quantum mechanics and Whiteheadian metaphysics, the *intensive valuation* of the subjective forms of alternative outcome states is closely linked with the logically prior, spatiotemporally *extensive coordination* of the related data, such that the vast majority of these data are coordinated and qualified as "environmental" to the subject occasion or system of occasions.

Put another way, the term "intensity" as Whitehead uses it refers to the *qualitative intensity* operative in genetic division, represented by the probability-valuated subjective forms/potential outcome states of the reduced density matrix in the supplementary stage/ mental pole. And this qualitative intensity is in part dependent upon a *quantitative intensity* which originates in coordinate division, operative in the primary stage/physical pole. The relationship between *coordinate extension* and *quantitative intensity* in the physical pole is evinced by the relationship between (i) the *number* of actualities extensively coordinated as "environmental" to the subject system, and (ii) the process of decoherence and transmutation in the supplementary stage. For apart from a sufficient number of actualities extensively coordinated as "environmental" to the subject system, there cannot be a sufficient number of negative prehensions to drive the process of decoherence/transmutation. Without this crucial negative selection process, there cannot be any subsequent qualitative valuation of intensities; there can be no reduction of the coherent superposition of the pure state density matrix to the decoherent set of probability-valuated, mutually exclusive and exhaustive potential outcome states/transmuted subjective forms in the reduced density matrix. Potential outcome states would instead remain locked in a coherent superposition; the data would persist as uncoordinated bare multiplicity, each datum superposed with all others, with some aspects of the superposition mutually consistent and others mutually contradictory. So long as the latter remain operative in the concrescence, there can be no reduction (transmutation) of the superposed multiplicity into a matrix of mutually exclusive and exhaustive, intensity-valuated transmutations. Thus the qualitative intensity operative in the mental pole cannot evolve apart from the extensive quantitative intensity that originates in the physical pole.

But the dipolarity of concrescence implies a conditioning in the opposite direction as well—a conditioning of the physical pole of a concrescence by the mental pole of an antecedent objectified actuality. Returning to the example of the salesman and his wife, consider answer (b), which emphasizes the extensive spatiotemporal coordination of the salesman-system and the wife-system. This coordination is relativistic, given the speed-of-light limitation of causal efficacy in the physical pole. Thus a speed-of-light communicative transmission from the wife to the salesman is the "fastest" way in which the salesman might be causally influenced by the birth event; for only events that fall within his past light cone are capable of causally influencing him. Whereas the EPR-type nonlocal causal phenomenon associated with the first answer was "causal affection of potentia by logically prior actuality"—a feature operative in the supplementary stage/mental pole and disclosed by the genetic analysis of concrescence—the causal phenomenon associated with answer

(b) is "causal influence of actualization by temporally prior actuality." It is a feature of concrescence operative in the primary stage/physical pole and disclosed by coordinate analysis. Since both salesman-system and wife-system are in the same general inertial reference frame relative to the Earth—that is, they are both relatively "stationary" on the Earth—the relativistic features of their spatiotemporal extensive coordinations do not pose much conceptual difficulty.

But suppose that the salesman were on a space transport on his way back to Earth from the Hong Kong Galaxy, twenty light years away, traveling at $0.8c$. (c = the speed of light, approximately 186,000 miles per second). The fact that c is a constant in all reference frames means that there will be variances between (i) the spatial extensive coordinations from the salesman's inertial reference frame, relative to (ii) the spatial extensive coordinations from the wife's inertial reference frame. This will be evinced in the phenomenon of "length contraction" from the perspective of the salesman's inertial reference frame, such that the distance in the direction of his travel between the Hong Kong Galaxy and Earth will be twelve light years from the perspective of his inertial reference frame, and twenty light years from the perspective of his wife's inertial reference frame. Similarly, there will be variances between (i) the temporal extensive coordinations from the salesman's inertial reference frame, relative to (ii) the temporal extensive coordinations from the wife's inertial reference frame. This will be evinced in the phenomenon of "time dilation" from the perspective of the salesman's reference frame, such that the time interval of his travel between the Hong Kong Galaxy and the Earth will be fifteen years relative to his inertial reference frame, and twenty-five years relative to his wife's inertial reference frame.

Coordinate Division and the Decomposition of Invariant Spacetime Intervals

These variances in spatial extensive coordinations alone, and temporal coordinations alone, reflect the relativity of "simultaneity." The synchronization of the wife's watch and the salesman's watch, for example, would vary depending upon the reference frame from which the synchronization were measured. But the causal relationships between *serial routes of events* associated with the salesman and with his wife—events whose extensiveness is coordinated both spatially *and* temporally together as events in spacetime constitutive of *spacetime intervals*—are *invariant* regardless of reference frame according to Einstein's special and general theories of relativity.

From the perspective of Whiteheadian metaphysics, this relativistic invariance of spacetime intervals and the associated objectivity of causally related *spatiotemporally coordinated extensive events* is the key feature of Einstein's theory of relativity—not the phenomenon of time dilation with respect to merely temporally coordinated events, or length contraction with respect to merely spatially coordinated events. These phenomena may be pronounced when one considers nature's most fundamental constituents to be material particles existing in three-dimensional space and moving about this space in time. But when one considers nature's most fundamental constituents to be quantum actual events occurring in four-dimensional spacetime, the invariance of spacetime intervals and the objectivity of the causally related spacetime events constitutive of them are the truly important features of spatiotemporal extension and relativity theory. Relativity theory, in other words, is most importantly a theory of the objective invariance of spacetime intervals and the relations among events constitutive of them. It is not merely a theory of the subjective variance of spatial coordination and temporal coordination. Commenting on relativity theory at a public lecture in 1908, the mathematician (and onetime instructor of Einstein) Hermann Minkowski proclaimed, "The views of space and time that I wish to lay before you have sprung from the soil of experimental physics, and therein lies their strength. They are radical. Henceforth space by itself, and time by itself, are doomed to fade away into mere shadows, and only a union of the two will preserve an independent reality."[3]

The key to relativistic invariance is, of course, the constancy of the velocity of light regardless of reference frame. Because of this constancy, light cones are invariant for every event in spacetime, and thus the causal relationships among events within one event's past light cone will hold universally, for any event in any reference frame. It may be that for a given event located at a particular set of spacetime coordinates, two causally, historically related spacetime events will lie in the past light cone of the percipient event. The salesman might, for example, spill a cup of coffee and burn his hand, both events being historically related events in his past light cone. Thus, these events can be—and very likely are—causally influential of the salesman at his spacetime coordinates. But these events might lie outside the past light cone of his wife, and as such, their causal relations cannot be causally influential of her. They might lie in her future light cone, or they might lie outside both light cones in the hazy realm of "her contemporary universe." But the logical, asymmetrical, historical ordering of those two events nevertheless holds universally. There can be no reference frame, for example, wherein the event of the coffee spill is subsequent to the salesman's hand being burned by the spill. And in the same way, there can be no reference frame wherein the salesman becomes a father prior to his child being born. Indeed, one can imagine some percipient event at a particular set of spacetime coordinates for whom all of the aforementioned events lie in its past light cone—the spilling of the coffee, the burning of the hand, and the birth of the child. The asymmetrical, logical, causal order of these spacetime events will be maintained for this percipient—in its case as a temporal order—for the same reason they are maintained for any extensive, spatiotemporally coordinated percipient event: because of the relativistic invariance of spacetime intervals.

This does not mean, however, that there must be some "absolute reference frame" according to which *all* actualities are *universally temporally ordered*, or *universally spatially ordered*. Extensive spatiotemporal coordination of events does indeed yield invariant spacetime intervals among these events; but such intervals are always only *partially* ordered: their asymmetrical causal order is globally invariant, but the various temporal and spatial coordinations of events in an invariant spacetime interval entail manifold diverse potential decompositions of the invariant spacetime interval into different sets of potential timelike and potential spacelike intervals of varying magnitudes. There might be, for example, many different potential timelike intervals which include two causally related events A and B. Some potential timelike intervals will entail time dilation, some more than others, and some none at all, and so all of these potential timelike intervals will entail different magnitudes. But the asymmetrical causal order of A and B remains universally invariant. Thus, spacetime events are limited to *partial ordering*, such that causal order among actualities always holds universally, but there will be many *potential spacelike intervals* and *potential timelike intervals* of different magnitudes spanning any two actual events. The salesman's journey from Earth to the Hong Kong Galaxy and back again, for example, entails event A—his leaving, and event B, his return. If his wife is present at both events, then given the effects of time dilation referred to above, the timelike interval between events A and B from the salesman's reference frame has a magnitude of thirty years (fifteen years out, and fifteen years back). Whereas the timelike interval between events A and B from the wife's reference frame is fifty years (twenty-five years out, and twenty-five years back). Regardless, though, the asymmetrical causal order among events associated with the salesman and the wife along these intervals is universal. The fact of time dilations does not, in other words, entail the possibility of a causal effect temporally preceding its cause. Causal agent-events will always temporally precede their causal patient-events, universally and regardless of time dilation or any other relativistic phenomenon.

Spacetime intervals, then, can be decomposed or coordinately divided into diverse potential mixtures of (i) "timelike" intervals of temporally related events (where one event lies within the light cone of the other event), (ii) "spacelike" intervals of spatially related events (where one event lies outside the light cone of the other event), and (iii) "lightlike" intervals of luminally related events (where one event lies on the light cone of the other event, such as events spanning an ideal electromagnetic transmission in space).

Thus, according to Einstein's relativity theory, the constancy of the velocity of light in all reference frames yields two types of relations among extensive spatiotemporally coordinated actualities: (i) symmetrical spatial and temporal relations among actualities belonging to different inertial coordinate systems (as regards, for example, the relativity of "simultaneity"); (ii) asymmetrical spatiotemporal relations among spacetime intervals as historical routes of actual events. The famous "twins paradox" is a popular thought experiment which exemplifies this distinction between symmetrical and asymmetrical extensive relations. It is similar to the salesman-wife example above, so we can incorporate them: Imagine twin sisters, one on Earth and one riding with our salesman on his starship, traveling at .8 c to the Hong Kong Galaxy twenty light years away. They arrive, turn around, and return to Earth, again traveling at 0.8c. Because of the effects of time dilation described earlier, when the twins are reunited after the trip, the traveling twin will be younger than her sister.

The reason this has occasionally been considered paradoxical within the framework of special relativity is because according to this theory, spatial and temporal relations among actualities belonging to different inertial coordinate systems, such as the Earth and the ship, are *symmetrical*. For from the perspective of the starship, it is *Earth* that is moving away from the ship as it heads toward the Hong Kong Galaxy; and from the perspective of Earth, it is the ship that is moving. Special relativity thus holds that there is no way to objectively distinguish between uniform motion and rest when relating inertial coordinate reference frames. Therefore, it is argued, there is no reason why *both* the earthbound sister and the traveling sister should not experience the effects of time dilation. Why is it that only the traveling sister has aged less as the result of the trip, and not the earthbound sister, too?

Within the framework of special relativity, which applies only to inertial reference frames, the answer is simple, though not too satisfying: The starship does not travel with uniform motion throughout the trip. It undergoes acceleration—changes speed as well as direction—when it turns around at the Hong Kong Galaxy and heads back home (and also presumably when it departs and returns to Earth, speeding up as it leaves and slowing down when it returns). A more satisfying answer, however, is found within the framework of general relativity, which applies to *all* spatiotemporal reference frames, including those undergoing acceleration: The moments of acceleration are historically, asymmetrically, causally related *events* in spacetime, in the same sense that the salesman's coffee spill and burned hand were such events in the previous example. These spatiotemporally coordinated events associated with the spacefaring twin form spacetime intervals that are asymmetrically related to other spacetime intervals, such as those intervals describing the earthbound twin. Associated with each twin, then, is a "worldline" comprising historical routes of extensive spatiotemporally coordinated events and the invariant spacetime intervals spanning these events. The earthbound twin's worldline is thus asymmetrically related to the traveling twin's worldline, as well as to the worldline describing the historical route of any other spatiotemporally extensive actuality.

Thus the logical historical ordering of event data in the mental pole finds its reflection in the relativistic spatiotemporally extensive coordinations of event data in the physical pole—so long as these data are *extensively coordinated spatiotemporally*. The relativity of spatiotemporal extensive coordination in the physical pole is not in any way at odds with the logical, historical relations of events given in genetic analysis of the mental pole; concrescence is dipolar, with the operations of each pole implicative of those of the other. There are indeed symmetrical relativistic *spatial coordinations* and *temporal coordinations* in the physical pole that stand opposed to the asymmetrical logical relations of data in the mental pole. This is to be expected, as the poles *are* diverse despite their mutual implication. But the *spatiotemporal* extensive coordination of actualities yields invariant spacetime intervals comprising asymmetrically ordered causal relations among the events constitutive of these intervals. These asymmetrical causal relations are universal and absolute.

Hence, not *every* extensive coordination in the physical pole is noninvariant relative to various reference frames. There are, in fact, many relativistic invariants—objective absolutes—operative in Einstein's theories of special and general relativity. Most important for this discussion is the invariance of the spacetime interval spanning two spatiotemporally coordinated events, such that their asymmetrical causal order is maintained universally. This might be seen as a reflection in the physical pole of the logical asymmetrical, historical ordering of data in the mental pole, and vice versa. But four-dimensional spatiotemporal extensive coordination in the physical pole yields other relativistic invariants as well, including momentum-energy, where energy is the temporal extensive aspect of the coordination and momentum the spatial extensive aspect; electric charge–current density is another, where charge density is the temporal extensive aspect of the coordination, and current density is the spatial extensive aspect.

There are, then, two fundamental interrelated principles underlying all of these relativistically invariant relations: (i) the constancy of the velocity of light; (ii) the four-dimensional spatiotemporal extensive coordination of event data in the primary stage of concrescence. Other coordinations are possible—spatial extensive coordinations, temporal extensive coordinations—and these will yield noninvariant spatial relations, and noninvariant temporal relations relative to diverse reference frames. But with respect to those aspects of extensive coordination of data in the physical pole which find their implication in the asymmetrical logical ordering of data in the mental pole, the key lies in the four-dimensional spatiotemporal extensive coordination of objectified data in the physical pole, productive of relativistically invariant intervals spanning events in spacetime.

EMPIRICAL ADEQUACY AND THE CONSTANCY OF THE VELOCITY OF LIGHT

It is clear that apart from constancy of the velocity of light, the invariance of spacetime intervals would be jeopardized. In this sense, this critical, constant velocity c provides a bridge between the logical order of actualities operative in the mental pole and the casual, spatiotemporal order operative in the physical pole. Were there velocities that exceeded c, there could be actualities existing simultaneously in numerous places, and the logical order and the directionality of time given by a closed past and an open future would be compromised. One could then posit, as Richard Feynman once playfully did, that there exists only one electron in the whole of the universe, zipping around in space and time from the past into the future and from the future into the past, such that whenever and wherever one observes an electron, one is really observing the same entity.

Thus, apart from the critical velocity c, the very identities of actualities would be as jeopardized as their logical order and their causal relations. And extensive spatial location would suffer the same fate as extensive temporal location: Nonlocal causal affection of potentia by logically prior actuality, of the sort exemplified by the EPR-type experiments, would be indistinguishable from local causal influence of actualization by temporally prior actuality. All notions of spatiotemporal extensive publicity and privacy would be lost.

But the constancy of the velocity of light is just one of the two fundamental principles from which Einstein deduced his theory of relativity—and the second is just as significant to Whiteheadian metaphysics as the first: *The laws of physics take the same form with respect to any inertial system of coordinates.*[4] Recall that Whitehead suggested four desiderata which any speculative metaphysical scheme must meet:[5] The scheme should be (i) *logical*; (ii) *coherent*, in the sense that its fundamental concepts are incapable of abstraction from each other; (iii) *empirically applicable*, meaning that the scheme must apply to at least some types of experience; and (iv) *empirically adequate*, in the sense that there are no types of experience conceivable that would be incapable of accommodation by the interpretation. The latter two desiderata (and indirectly the first) are clearly exemplified by the two principles from which Einstein deduced his theory of relativity: (i) The laws of physics take the same form with respect to any inertial system of coordinates; (ii) the speed of light is c with respect to any inertial system of coordinates.

As regards this latter principle, it should be emphasized that for Whitehead (and likely as well for Einstein) the critical importance of the constant c had little to do with its relation to the phenomenon of *light per se*; its significance, rather, lay in the derivative invariance of spacetime intervals and the associated possibility of (i) the asymmetrical, logical and causal ordering of events within spacetime reference frames, and (ii) the provision of a congruence relation that allows for the comparison of spatial and temporal extensive coordinations across diverse spacetime reference frames. In other words, the only reason that spacelike intervals and timelike intervals are *relatable* at all among diverse spacetime reference frames is that the velocity of light is constant in all frames. Indeed, for Whitehead, the speed of light in a vacuum was likely but a specific approximation of a generic critical velocity c operative in relativity theory:

> The critical velocity c which occurs in these formulae has now no connexion whatever with light or with any other fact of the physical field (in distinction from the extensional structure of events). It simply marks the fact that our congruence determination embraces both times and spaces in one universal system, and therefore if two arbitrary units are chosen, one for all spaces and one for all times, their ratio will be a velocity which is a fundamental property of nature expressing the fact that times and spaces are really comparable.[6]

> Both light and sound are waves of disturbance in the physical characters of events; and the actual course of the light is of no more importance for perception than is the actual course of the sound. To base the whole philosophy of nature upon light is a baseless assumption.

The Michelson-Morley and analogous experiments show that within the limits of our inexactitude of observation the velocity of light is an approximation to the critical velocity "c" which expresses the relation between our space and time units.[7]

Coordinate Extensive Relations and Genetic Historical Relations: The Dipolarity of Relativity Theory and Quantum Mechanics

The dipolar relationship between (i) the spatiotemporally extensive, relativistic, coordinate analysis of concrescence, and (ii) the historically extensive, logical, genetic analysis of concrescence, entails an analogous relationship between relativity theory and quantum mechanics, respectively. At the heart of this relationship is the identification of invariant spacetime intervals with the quantum mechanical state evolution of the events comprised by these intervals. A null spacetime interval, then, would represent a single quantum state. The interval spanning the emission and absorption of a photon is such a null spacetime interval—which would imply that photon emission-absorption is, from the standpoint of both coordinate and genetic division, a single quantum event. Given this, and the invariance of c, one could characterize a photon of electromagnetic energy as a quantum interaction between two events along a null spacetime interval—that is, a concrescence whose subject and physical prehensions span this null spacetime interval. For Whitehead, in reference to the above two quotations, this would be preferable to the characterization of a photon as a luminous, "mass-less particle" which "travels" at the constant velocity c in a "vacuum." All three quoted terms imply a fundamental mechanistic-materialistic conception of nature.

The decomposition of an invariant spacetime interval yields relativistically noninvariant timelike and spacelike intervals. Actualities that lie in the past light cone of a given event—that is, actualities constitutive of a timelike interval—are logically and historically ordered, such that regardless of reference frame, local asymmetrical causal orders hold globally; but again, this constitutes only a *partial* extensive ordering of events, not a global ordering, for there are manifold potential spacelike and timelike intervals of varying magnitudes which might span any two events. As regards the example of the twins, two asymmetrically related events, "ship leaving Earth" and "ship returning to Earth" entail an invariant spacetime interval which can be decomposed into different potential spacelike and timelike intervals of varying magnitudes. The timelike interval associated with the spacefaring twin, for example, is dilated relative to that of the earthbound twin, even though these intervals span the same two events.

Put another way, the limitation of partial ordering among actualities in spacetime ultimately derives from the fact that not all events are merely timelike separated. There are spacelike separations to be considered as well. An event C might lie outside the past light cone of causally related events A and B (i.e., where A lies in the past light cone of B); thus C will be spacelike separated from, and essentially contemporaneous with *both* A and B. Nevertheless, the logical, asymmetrical order of A and B holds universally—*even for C*, despite the causal limitations inherent in its spacelike separation from A and B. This brings us back to the example of the salesman and his wife. Event C above is associated with the salesman; event A is associated with his wife just prior to the birth, and event B is associated with the birth. We saw earlier that quantum mechanically, event C is in fact logically and historically asymmetrically related to events A and B, despite the fact that in terms of spatiotemporal extension, the relations between C and A, and C and B are symmetrical. Again, this symmetry derives from the fact that spatiotemporal extensive coordination yields only a partial ordering among the events coordinated. The EPR-type experiments demonstrate this same distinction between (i) global, asymmetrical, logical-historical quantum mechanical relations among *spacelike* separated events, and (ii) local, asymmetrical, temporal-historical extensive relations among *timelike* separated events. Causal affection of potentia by logically prior actuality pertains to the former; and causal influence of actualization by temporally prior actuality pertains to the latter.

In some sense, then, the partial spatiotemporal ordering of events constitutive of invariant spacetime intervals is never exclusive of—that is, never merely complementary to—the global quantum mechanical historical ordering of events. If the significance to experience of *spatiotemporally extensive coordinate relations among actualities* is to maintain its reflection in modern physics, and if the significance to experience of *historically extensive genetic relations among potentia and actualities* is to *find* its reflection in modern physics, then the rapprochement of relativity theory and quantum theory would seem to require a dipolar relationship of the sort given in Whiteheadian metaphysics. The alternatives are either to reduce extensive, relativistic, coordinate relations to logical, quantum mechanical, genetic relations, or to do the reverse. The first alternative, sometimes associated with information theoretic quantum mechanical cosmologies, vitiates the distinctiveness of local, spatiotemporally extensive relations among actualities and the relativistic "privacy" of such relations, given the possibility of their spacelike extensive separation from other locally related actualities. The second alternative, associated with the "hidden variables" approaches to quantum mechanics, vitiates the distinctiveness of nonlocal relations among potentia and actualities, and the historical "publicity" of such relations given the ontological significance of potentia and the possibility that every history of actualities is related to a broader history of actualities subsuming it—and that every *potential* history of actualities is so related.

Such reductions entail gross eliminations of important aspects of experience. Since physics is largely rooted in inductions from experience, such reductions are clearly arbitrary. If they are to be avoided, the rapprochement of relativity theory and quantum mechanics would seem to require a dipolar metaphysical theory, whereby relativistic, spatiotemporally extensive coordinate division, and quantum mechanical, historically extensive genetic division are mutually implicative.

By such a dipolar approach, one might correlate the global-historical, asymmetrical ordering of actualities operative in quantum mechanics with the local-temporal, partial asymmetrical ordering of relativity theory. Recall that in the latter, invariant spacetime intervals among actualities—that is, *actual relations among actualities*—are decomposed into *potential relations among actualities*—that is, there is coordinate division of *actual* invariant spacetime intervals into manifold alternative sets of relativistically noninvariant, *potential* spacelike and timelike intervals with diverse magnitudes. Thus, the dipolar approach also implies a direct relationship between such *coordinate extensive indeterminacy* given by the theory of relativity and the *genetic quantum indeterminacy* given by the Heisenberg uncertainty relations. The speed of light operative in the former, in other words, is closely related to Planck's constant in the latter. Recall, for example, that according to the Heisenberg uncertainty relations in quantum mechanics, the uncertainty of the measured energy of a system over a time interval Δt is always at least $h/4\pi$:

$$\Delta E \Delta t \geq \frac{h}{4\pi}$$

where h is Planck's constant.

And according to the special theory of relativity, the uncertainty of the measured time difference Δt between two events separated by a spatial distance Δx requires that $\Delta t/\Delta x$ is always at least $1/c$. Put a different way, the Heisenberg uncertainty relations hold that it is impossible to simultaneously specify the *momentum* and *position* of a particle; and special relativity holds that it is impossible to determine whether an inertial reference frame is "*truly moving uniformly*," or "*truly at rest.*"

Recall that the "hidden variables" answer to quantum indeterminacy entailed the hypothesis that all measured events are deterministically interrelated by their entanglement with an "ether" of unmeasured, *globally extensively ordered*, "hidden" actualities. Were these hidden actualities disclosed and incorporated into the measurement, the indeterminacy given by $h/4\pi$ would disappear. There was a similar "hidden variables" approach to special relativity, represented by the thinking of physicist Hendrik Lorentz, who had developed the formulae describing the phenomena of time dilation, length contraction, and mass increase—the "Lorentz transformations"—foundational to Einstein's special theory of relativity. Lorentz proposed the existence of an "absolute" spacetime reference frame that would provide for a total deterministic temporal ordering of events. Thus one could explain away the indeterminacy, on the scale of $1/c$, of the decomposition of an invariant spacetime interval of logically ordered events into indeterminate, noninvariant, potential spacelike and timelike intervals. Against Lorentz's notion of an absolute spacetime reference frame, Einstein insisted upon the indeterminacy of the decomposition of spacetime intervals, even though he argued against the indeterminacy of quantum mechanical measurement interactions; for some reason, in the case of the latter, the hidden-variables approach was the more reasonable to him.

THE ISSUE OF ABSOLUTE SPACETIME AND ALTERNATIVE CONCEPTIONS

In his earlier writings on relativity, Whitehead himself maintained an idea, associated with that of Lorentz, that there must be a uniform, absolute metrical structure to spacetime that is wholly independent of events relativistically coordinated by diverse spacetime reference frames. In *Process and Reality*, however, he refined his position, such that the metrical structure of spacetime was not entirely independent of events and their relativistic extensive coordinations—a refinement consistent with the experimental confirmations of general relativity that were produced during that time. But the *abstract extensive morphological elements* of spacetime—including geometrical elements such as points, lines, and planes—must be independent of events; for it is the invariance of these abstract extensive morphological elements that provides the congruence relations necessary for the comparison of spatial and temporal extensive coordinations across diverse spatiotemporal reference frames. Whitehead fully develops his theory of extensive abstraction as a theory of "extensive connection" in Part IV of *Process and Reality*, and it will be explored in more detail later in this chapter.

It is understandable why Whitehead would have initially sided with the Lorentzian characterization of spacetime as fundamentally absolute and only abstractly relativistic. The desire was to reconcile two conflicting yet empirically demonstrable features of spacetime indicated by the special theory of relativity: (i) the equivalence of all inertial reference frames; and (ii) the ontological significance of spacetime events, such as changes in velocity and direction—that is, acceleration events. Whitehead commented on these conflicting features as exemplified by the "twins paradox" in an essay presented at a 1923 Aristotelian Society symposium entitled "The Problem of Simultaneity: Is There a Paradox in the Principle of Relativity in Regard to the Relation of Time Measured to Time Lived?"[8]

Again, Whitehead focuses on the apparent conflict in the special theory of relativity between: (i) the equivalence of all inertial reference frames—those undergoing "uniform motion" and those "at rest"—implying that these characterizations lack any ontological significance. The characterizations "truly in uniform motion" and "truly at rest" are, in other words, merely mutually implicative epistemic characterizations; and (ii) the ontological significance of spacetime *events*, such as *changes* in velocity—that is, acceleration events. Spacetime events and their relations are ontologically significant because the spacetime intervals spanning such events—the relations between events—are *globally* invariant across diverse *local* reference frames.

If the relativistic features of spacetime are of epistemic importance only, and the ontological features are only those which are absolute, then it is understandable why Whitehead might conclude that it is the latter which should set the fundamental shape of our metaphysical and cosmological theories of spacetime. But there are three incompatible deductions one can make from this initial premise of the absoluteness of spacetime: The absoluteness of events and their objective relations—represented by the invariance of spacetime intervals between events—implies (i) a relation between these events and a fundamentally uniform, absolute spacetime; or (ii) a relation between these events and a fundamentally uniform, absolute, abstract morphological structure underlying spacetime—that is, a universally unique abstract system of reference; or (iii) that the absolute events and their objective relations, represented by the invariance of spacetime intervals, *constitute* spacetime itself.

Einstein resolved the aforementioned conflict inherent in special relativity by proposing his general theory of relativity—essentially a field theory of spacetime that would encompass all reference frames, not just inertial frames. Einstein's intention was that his general theory of relativity would exemplify deduction (iii) above. Spacetime, for Einstein, *is* actual events and their relations. Thus spacetime presupposes actualities and their relations, as does measurement of spacetime, such that spacetime essentially presupposes measurement. In *Process and Reality*, Whitehead proceeds from deduction (ii) when presenting his theory of relativity. For him, the morphological abstractive forms by which we order actualities both spatially and temporally must maintain their existence independent of these actualities if there are to be any relations of congruence among diverse spacetime systems—relations such as those given in the Lorentz transformations.

The contrast of Whitehead's and Einstein's positions is clearly exemplified in the concept of a "straight segment" as given by each position. According to the general theory of relativity, which entails a curved, non-Euclidean spacetime geometry (the curvature being a representation of gravity), a "straight segment" is the "straightest possible path" between two events in spacetime (i.e., a geodesic). Thus, "straightness" presupposes events and their relations via measurement. Whitehead suggests that the toleration of this presupposition has its roots in the Euclidean definition of "straight line" which is similarly predicated upon "points" and their morphological relations. And the concept of "point" itself as given by Euclid seems to entail or at least imply the notion of a special class of ultimate physical entity:[9] "A straight line," according to the fourth definition of Euclid's *Elements*, "is any line which lies *evenly* with the *points* on itself."[10] Whitehead remarks:

> The weak point of the Euclidean definition of a straight line is, that nothing has been deduced from it. The notion expressed by the phrases "evenly," or "evenly placed," requires definition. The definition should be such that the uniqueness of the straight segment between two points can be deduced from it. Neither of these demands has ever been satisfied, with the result that in modern times the notion of "straightness" has been based on that of measurement. A straight line has, in modern times, been defined as the shortest distance between two points. In the classic geometry, the converse procedure was adopted, and measurement presupposed straight lines. But, with the modern definition, the notion of the "shortest distance" in its turn requires explanation.[11]

Indeed, in the curved geometry of spacetime given in general relativity, the "straightest possible path" between two spacetime events is not necessarily the "shortest possible path"; moreover, there can be more than one "straightest possible path," so the definition must be even *further* qualified by the entities related: The "straightest possible path" in the curved spacetime of general relativity is a geodesic, defined as the path along which the magnitude of the timelike interval between two events is greatest when compared to any other interval connecting the events. Thus the straightest possible path between two spacetime events is the path of a freefall reference frame—a reference frame undergoing (ideally) no acceleration. With respect to the "twins paradox" discussed and resolved earlier (and assuming the twins are united at the beginning and end of the journey), the "straightest possible path" through spacetime which includes the event of the ship's leaving Earth and the event of its return would be the worldline of the earthbound sister. Her worldline, in other words, follows a geodesic connecting these events (ignoring, for the sake of simplicity, the gravitational, rotational, and orbital accelerations associated with her earthly reference frame).

Thus the concept of "straight segment between two points" as "straightest possible path between two spacetime events" as given in general relativity clearly does not satisfy Whitehead's requirement that the *unique* abstract extensive notion of "straight segment between two points" can be deduced solely from the definitions of those other unique abstract extensive notions it presupposes. Whitehead's theory of extensive abstraction given in Part IV of *Process and Reality* as his "theory of extensive connection" and discussed more fully later in this chapter, is intended to satisfy these requirements.

As regards general relativity, Einstein's embrace of the premise that absolute events and their objective relations *constitute* spacetime itself was his attempt to satisfy the logical positivist complaints about the synthetic a priori concept of absolute spacetime. The physicist and philosopher Ernst Mach, for example, vehemently repudiated the idea that an immutable "absolute spacetime" could in some way be ingredient in or affective of actual events without these events being reciprocally ingredient and affective of spacetime. In his view, such arbitrarily sanctioned asymmetry was intolerable. The desire, of course, was to do away with the concept of a physically affective "spacetime" altogether. But Einstein's theory of general relativity alleviated the asymmetry by doing the very opposite, characterizing "absolute" spacetime as nothing other than "absolute" events and their "absolute" relations.

But the premise from which Whitehead starts—that the absoluteness of events and their objective relations implies a relation between (i) actual events, and (ii) a fundamentally uniform, absolute morphological spacetime structure or abstract system of reference—is not wholly absent from Einstein's general theory of relativity. For this theory cannot be put into use without the imposition of global boundary/symmetry conditions—that is, a preferred, abstract coordinate system of reference. As the general theory of relativity is applied to cosmological models, for example, the cosmic background radiation generated by the "big bang" often corresponds to a particular set of spacetime intervals that yields an isotropic distribution of the background radiation. A related concept is the ratio of the energy density of the universe to a given "critical density" determined by its expansion rate. Different values of this ratio, Ω, correspond to different spatial universal geometries: $\Omega > 1$ corresponds to a closed universe of finite volume with positive (spherical) curvature; $\Omega < 1$ corresponds to an open universe of infinite volume with negative (hyperbolic) curvature; and $\Omega = 1$ corresponds to a flat Euclidean geometrical universe of infinite volume. Each of these alternatives entails related sets of possible symmetry or boundary conditions, amounting to a unique abstract coordinate system of reference that must be adopted in the practical application of the theory.

General relativity, then, is not wholly compatible with the premise, "Absolute events and their objective relations, represented by the invariance of spacetime intervals, *constitute* spacetime itself." Rather, general relativity entails that spacetime is the product of the objective relations among absolute events *extensively coordinated by* and *related to* some absolute background spacetime metric. This spacetime metric is an abstract extensive element of the sort Whitehead proposes in *Process and Reality*; it is not determined by the events and their relations, though it is related to them. "The concrescence presupposes its basic region," writes Whitehead, "and not the region its concrescence."[12]

Returning again to the "twins paradox": One cannot spatiotemporally distinguish between the sisters solely in terms of their mutual relations; for their individual reference frames, when so compared, are purely symmetrical. One cannot, in other words, simply compare their individual worldlines in spacetime; one must, rather, compare these worldlines as coordinated by an abstract extensive universal spacetime metric. Apart from this metric, there can be no congruence relations between the two worldlines—no basis for the comparison of the decomposition of each spacetime interval into its respective spacelike and timelike coordinations; a time-dilated "year" as *locally experienced* by the traveling sister would have no correlation with a "year" as *locally experienced* by the earthbound sister. Apart from coordination by an abstract extensive universal

spacetime metric, there could be, in other words, no objective ontological comparison of the two worldlines whatsoever; their comparison would remain forever symmetrical and thus merely epistemically
significant.

The Dipolar Cooperation of Coordinate and Genetic Division

In summary of the discussion thus far, by the dipolarity of concrescence, the logical, asymmetrical, historical relations among data in
the mental pole—those that are of particular importance to the decoherence interpretations of quantum mechanics—are reflected in
the relativistically invariant asymmetrical relations among extensive,
spatiotemporally coordinated data in the physical pole. These data
take the form of invariant spacetime intervals, constitutive of actual
occasions and their extensive connections. Also, the *qualitative intensities* yielded in the mental pole derive in part from the *quantitative
intensities* yielded in the physical pole; these quantitative intensities
pertain to the spatiotemporal coordination of data into an extensive
region of environmental actualities whose number is directly related
to the process of negative selection/negative prehension in the mental pole. Apart from a quantitative intensity of environmental extension and entanglement with the subject-object system—a function of
the number of actualities extensively coordinated as "environmental"
to the subject-object system—decoherence and transmutation cannot occur, and the qualitative, valued intensities operative in the
mental pole will not be yielded. Potentia will thus persist as unintegrated sheer multiplicity, and potential outcome states will remain
locked in a coherent superposition.

Any theory of physics intended to be accommodated by Whitehead's metaphysical scheme must give ontological significance to the
extensive spatiotemporal coordination and connection of actualities.
To elevate to primacy the logical historical relations among actualities and potentia—which, given the EPR-type nonlocality experiments is but a single feature of nature experimentally evinced by
quantum mechanics, among so many other features—is to disregard
the fundamental dipolarity of concrescence, and thereby disregard
the entire metaphysical scheme. Indeed, the relegation of spatiotemporal extensive coordination to the realm of mere subjective abstrac-

tion is ironically reminiscent of the relegation of potentia to the
realm of mere abstraction typical of mechanistic materialism. Apart
from the dipolarity of concrescence, the scheme loses all coherence.
The operations of the mental pole presuppose the extensive, spatiotemporal, coordinative operations of the physical pole, and thus cannot be elevated above those operations. Whitehead writes:

> Extensive connection with its various characteristics is the fundamen-
> tal organic relationship whereby the physical world is properly de-
> scribed as a community. There are no important physical relationships
> outside the extensive scheme. To be an actual occasion in the physical
> world means that the entity in question is a relatum in this scheme of
> extensive connection. In this epoch, the scheme defines what is physi-
> cally actual.[13]

The genetic division of a concrescence operative in the mental
pole emphasizes the quantum singularity of the subject and the
unity of its logically ordered world of antecedent actualities. This
logically ordered, historical world finds its reflection in the invariant
spacetime intervals among data operative in the coordinate division
of a concrescence in the physical pole. Whitehead writes, "The subjective unity of the concrescence is irrelevant to the divisibility of
the region. In dividing the region we are ignoring the subjective unity
which is inconsistent with such division. But the region is, after all,
divisible, although in the genetic growth it is undivided."[14] But because each subject occasion exists in its own particular reference
frame, with its own particular standpoint, coordinate division for
that subject will entail a propositional coordinate division (or "decomposition" in the terminology of relativity theory) of the invariant
spacetime intervals of the actual world into relativistically noninvariant, potential spacelike and timelike intervals *for* that subject, in *its*
actual world. As regards the "twins paradox," for example, the invariant spacetime worldline of the spacefaring sister as she heads toward
Earth will be coordinately decomposed into a potential *dilated* timelike interval and potential *length-contracted* spacelike interval *relative
to* the coordinate division of the earthbound sister's spacetime
worldline. Whitehead writes:

> So this divisible character of the undivided region is reflected into the
> character of the satisfaction. When we divide the satisfaction coordi-
> nately, we do not find feelings which *are* separate, but feelings which

might be separate. In the same way, the divisions of the region are not divisions which *are*; they are divisions which *might be*. Each such mode of division of the extensive region yields "extensive quanta": also an "extensive quantum" has been termed a "standpoint."[15]

The fact of invariant spacetime intervals spanning the extensive continuum of all actualities serves as a bridge between (i) the coordinate division of *actual* invariant spacetime intervals into relativistically standpoint-dependent, objectified, *potential* spacelike and timelike intervals in the physical pole, and (ii) the genetic division of the *actual* continuum into unified, logically ordered, *potential* historical routes of occasions. Thus the coordinate divisions of spacetime are propositional in that they yield *standpoint-dependent*, relativistically noninvariant, potential, coordinate spacelike and timelike intervals. This fact has its reflection in the multiple valued propositional transmutations of data—that is, potential unified quantum histories valued as probabilities—in the mental pole. Thus the valued potential subjective forms or transmuted propositional *unifications* of data in the mental pole are correlated with relativistic, propositional coordinate *divisions* in the physical pole. Sometimes data in the physical pole are objectified according to these extensive *physical* coordinate divisions; and sometimes these data are objectified by potentia operative in the mental pole—those pertaining to alternative probability-valued outcome histories that would correlate with *unrealized* potential coordinate decompositions of invariant spacetime intervals into relativistic spacelike and timelike intervals in the physical pole. These unrealized potential intervals, in other words, are coordinate divisions that differ from those initially produced in the physical pole according to the subject's standpoint; they are divisions that might have been objectified but were not.

> Thus the subjective form of this coordinate division is derived from the origination of conceptual feelings which have regard to the complete actual entity, and not to the coordinate division in question. Thus the whole course of the genetic derivation of the coordinate division is not explicable by reference to the categoreal conditions governing the concrescence of feeling arising from the mere physical feeling of the restricted objective datum. The originative energy of the mental pole constitutes the urge whereby its conceptual prehensions

adjust and readjust subjective forms and thereby determine the specific modes of integration terminating in the "satisfaction."[16]

Whitehead suggests three doctrines that reflect the relationship between (i) the standpoint-dependent coordinate divisions, in the physical pole, of the invariant spacetime intervals defining the actual world as extensive continuum; and (ii) the actual world as unified, logically and historically ordered society by genetic division in the mental pole:

> First, there is the doctrine of "the actual world" as receiving its definition from the immediate concrescent actuality in question. Each actuality arises out of its own peculiar actual world. Secondly, there is the doctrine of each actual world as a "medium." According to this doctrine, if S be the concrescent subject in question, and A and B be two actual entities in its actual world, then either A is in the actual world of B, or B is in the actual world of A, or A and B are contemporaries. If, for example, A be in the actual world of B, then for the immediate subject S there are (1) the direct objectification of A in S, and (2) the indirect objectification by reason of the chain of objectification, A in B and B in S. Such chains can be extended to any length by the inclusion of many intermediate actualities between A and S.
>
> Thirdly, it is to be noticed that "decided" conditions are never such as to banish freedom. They only qualify it. There is always a contingency left open for immediate decision. This consideration is exemplified by an indetermination respecting "*the* actual world" which is to decide the conditions for an immediately novel concrescence. There are alternatives as to its determination, which are left over for immediate decision. Some actual entities may be *either* in the settled past, *or* in the contemporary nexus, or even left to the undecided future, according to immediate decision. Also the indirect chains of successive objectifications will be modified according to such choice. These alternatives are represented by the indecision as to the particular quantum of extension to be chosen for the basis of the novel concrescence.[17]

Alternative potential coordinate divisions in the physical pole have their analogs in alternative potential genetic divisions or histories in the mental pole; these are the probability-valued, decoherent, potential quantum mechanical outcome states, or valuated propositional transmutations, which evolve in the supplementary stage of concrescence. These transmutations entail a process of negative selection or elimination via negative prehension of those actualities excluded from the particular standpoint-dependent coordinate decomposition effected in the physical pole:

> A coordinate division is thus to be classed as a generic contrast. The two components of the contrast are, (i) the parent actual entity, and (ii) the proposition which is the potentiality of that superject having arisen from the physical standpoint of the restricted subregion [i.e., the relativistically noninvariant subregion of spacelike and timelike intervals yielded by the propositional coordinate decomposition/ division of the invariant spacetime intervals of the whole extensive region.] The proposition is thus the potentiality of eliminating from the physical pole of the parent entity all the objectified actual world, except those elements derivable from that standpoint; and yet retaining the relevant elements of the subjective form.[18]

In the mental pole, the coherent superposition of indefinitely divisible states of the pure state density matrix, prior to decoherence and transmutation, have yet to be *qualified* by the coordinate decomposition yielded in the physical pole. This qualification occurs during the process of decoherence and transmutation, and entails the related processes of negative selection described previously. The result is a reduced density matrix of qualified, probability-valuated, alternative "true" propositions—that is, probability-valuated alternative potential quantum mechanical outcome states:

> The unqualified proposition is false, because the mental pole, which is in fact operative, would not be the mental pole under the hypothesis of the proposition. But for many purposes, the falsity of the proposition is irrelevant. [I.e., when relativistic variances in the decomposed spacetime intervals are insignificant from the subject's standpoint. Correlating the spatiotemporal experiences of the twin sisters discussed above, for example, would *not* be one of these cases—nor would a nonlocal EPR-type experiment.] The proposition is very complex; and with the relevant qualifications depending on the topic in question, it expresses the truth. In other words, the unqualified false proposition is a matrix from which an indefinite number of true qualified propositions can be derived. The requisite qualification depends on the special topic in question, and expresses the limits of the application of the unqualified proposition relevantly to that topic.
>
> The unqualified proposition expresses the indefinite divisibility of the actual world; the qualifications express the features of the world

[i.e., the spatiotemporally extensive features] which are lost sight of by the unguarded use of this principle. The actual world is atomic, but in some senses it is indefinitely divisible.[19]

The principle by which one is able to correlate (i) the operations of the physical pole, wherein the extensive continuum of the actual world comprises manifold invariant potential spacetime intervals among actualities, with (ii) the operations of the mental pole, by which the world is a unified, logically and historically ordered nexus of actualities, is the same principle by which relations are possible *within* the physical pole—that is, relations among diverse spacelike and timelike decompositions of invariant spacetime intervals. It is the principle of extensive abstraction. This principle evinces an abstract extensive order among diverse, relativistically noninvariant potential decompositions of invariant spacetime intervals—an order that provides the congruence relations by which these diverse decompositions are comparable; but it also evinces an abstract extensive order between these noninvariant *potential* decompositions and the invariant *actual* spacetime intervals from which they are decomposed; and also between these invariant and noninvariant extensive coordinations in the physical pole, and the world as a unified, logically and historically ordered nexus in the mental pole:

> The atomic actual entities individually express the genetic unity of the universe. The world expands through recurrent unifications of itself, each, by the addition of itself, automatically recreating the multiplicity anew.
>
> The other type of indefinite multiplicity, introduced by the indefinite coordinate divisibility of each atomic actuality, seems to show that, at least for certain purposes, the actual world is to be conceived as a mere indefinite multiplicity.
>
> But this conclusion is to be limited by the principle of "extensive order" which steps in. The atomic unity of the world, expressed by a multiplicity of atoms, is now replaced by the solidarity of the extensive continuum. This solidarity embraces not only the coordinate divisions within each atomic actuality, but also exhibits the coordinate divisions of all atomic actualities from each other in one scheme of relationship. . . .
>
> This scheme of *external* extensive relationships links itself with the schemes of internal division which are *internal* to the several actual entities. There is, in this way, one basic scheme of extensive connec-

tion which expresses on one uniform plan (i) the general conditions to which the bonds, uniting the atomic actualities into a nexus, conform, and (ii) the general conditions to which the bonds, uniting the infinite number of coordinate subdivisions of the satisfaction of any actual entity, conform.[20]

These extensive relations do not make determinate *what* is transmitted; but they do determine conditions to which all transmission must conform. They represent the systematic scheme which is involved in the real potentiality from which every actual occasion *arises*. This scheme is also involved in the attained fact which every actual occasion *is*. The "extensive" scheme is nothing else than the generic morphology of the internal relations which bind the actual occasions into a nexus, and which bind the prehensions of any one actual occasion into a unity, coordinately divisible.[21]

It is vital that this extensive order maintain its independence from the actualities ordered—that is, that it remain abstractive. Returning for a moment to our earlier discussion of the abstract concept of "straight line" defined nonabstractively in general relativity: A "straight segment" is there defined as the "straightest possible path between two spacetime events," which must be even further qualified by actual events as "a geodesic, which is the path along which the magnitude of the timelike interval between two events is greatest when compared to any other interval connecting the events." This amounts to a deduction of spatiotemporal extensiveness from the notions of actual "part" and actual "whole"—and by the principle of relativity, the "whole" of one spacetime interval or worldline is not the same as the "whole" of another, even if they include common actual events.

Though Whitehead's earlier theory of extensive abstraction was in fact derived from such notions of "part" and "whole," the theory as developed in *Process and Reality* defines abstract extensive elements such as "point," "straight line," and "flat loci" without presupposing actualities or relations among actualities—that is, measurement:

> Extensiveness is the pervading generic form to which the morphological structures of the organisms of the world conform. These organisms are of two types: one type consists of the individual actual entities; the other type consists of nexūs of actual entities. Both types are correlated by their common extensiveness. If we confine our attention to

the subdivision of an actual entity into coordinate parts, we shall conceive of extensiveness as purely derived from the notion of "whole and part," that is to say, "extensive whole and extensive part." This was the view taken by me in my two earlier investigations of the subject [see Whitehead's *The Principles of Natural Knowledge* and *The Concept of Nature*]. This defect of starting point revenged itself in the fact that the "method of extensive abstraction" developed in those works was unable to define a "point" without the intervention of the theory of "duration." [A "duration" is a nexus of contemporary actualities "in unison of becoming."] Thus what should have been a property of "durations" became the definition of a point. . . .

This whole question is investigated in the succeeding chapters of [Part IV]. Also I there give a definition of a straight line, and of "flat" loci generally, in terms of purely extensive principles without reference to measurement or to durations.[22]

It cannot be stressed enough that spatiotemporal extension is of fundamental importance to Whiteheadian metaphysics. The latter simply cannot coherently accommodate, for example, interpretations of quantum mechanics which imply a reduction of nature to sheerly logically ordered and fundamentally disembodied "information" from which physical extension is but an abstraction. Indeed, such interpretations and associated misinterpretations of Whitehead's thought have little compatibility with the principles and spirit of speculative philosophy foundational to his metaphysics; they are fueled, rather, by the spirit of sheer reductionism—the same spirit that inspired the equally narrow worldview of fundamental materialism wholly disavowed by Whitehead's philosophy of organism. Physical time and physical space are, for Whitehead, as conceptually significant as the concept of "creative advance," and all three concepts equally presuppose the concept of extension:

> In this general description of the states of extension, nothing has been said about physical time or physical space, or of the more general notion of creative advance. These are notions which presuppose the more general relationship of extension. They express additional facts about the actual occasions. The extensiveness of space is really the spatialization of extension; and the extensiveness of time is really the temporalization of extension. Physical time expresses the reflection of genetic divisibility into coordinate divisibility. . . .
>
> The immediately relevant point to notice is that time and space are characteristics of nature which presuppose the scheme of extension.

But extension does not in itself determine the special facts which are true respecting physical time and physical space.[23]

THE DIPOLARITY OF PUBLICITY AND PRIVACY

It was mentioned earlier that all conceptions of spatiotemporally "regional" or "private" locality have their foundation in the constancy of the critical velocity c and its operation in coordinate division. Apart from this critical velocity, the invariance of spacetime intervals and thus their asymmetrical logical order would be jeopardized. By the critical velocity c, coordinate decompositions of these invariant intervals are never able to violate the logical asymmetry of causal relations, from antecedent cause to subsequent effect. Apart from the constancy of the critical velocity c, this one-way direction of causality would degenerate into symmetry; the past would cease to remain closed, and the future would cease to remain open. Relations among all entities, local and nonlocal, would be potentially maximally intimate. The coordinate privacy of local causal influence of actualization by temporally prior actuality, and the genetic publicity of nonlocal causal affection of potentia by logically prior actuality, would be whitewashed by a global "causal influence by antecedent and subsequent actuality"; indeed, the very qualifications "antecedent" and "subsequent" would cease to have any universal meaning.

In Whitehead's metaphysical scheme, notions of publicity and privacy are intimately related to the actual occasion as "subject-superject" in concrescence. The occasion as actualizing subject is a quantum of private immediacy—an entity undergoing its own actualization; and the occasion as actualized superject serves as a public datum to be prehended and objectified by subsequent concrescences:

> In the analysis of actuality the antithesis between publicity and privacy obtrudes itself at every stage. There are elements only to be understood by reference to what is beyond the fact in question; and there are elements expressive of the immediate, private, personal, individuality of the fact in question. The former elements express the publicity of the world; the latter elements express the privacy of the individual.
>
> An actual entity considered in reference to the publicity of things is a "superject"; namely, it arises from the publicity which it finds,

and it adds itself to the publicity which it transmits. It is a moment of passage from decided public facts to a novel public fact. Public facts are, in their nature, coordinate.

An actual entity considered in reference to the privacy of things is a "subject"; namely, it is a moment of the genesis of self-enjoyment. It consists of a purposed self-creation out of materials which are at hand in virtue of their publicity.[24]

Thus, in addition to the notions of privacy and publicity relative to the actual occasion as "subject-superject," there are notions of privacy and publicity relative to the concrescence as a dipolar entity, with public external relations and publicly influenced coordinations in the physical pole, and private internal relations in the mental pole, where the public data from the physical pole are internally and privately integrated and valuated. These dipolar aspects of publicity and privacy pertain not just to the actualizing subject, but also to the spatiotemporally coordinated superject actualities that serve as prehended and objectified data in the physical pole of a subsequent concrescence. These data enjoy a certain privacy in the physical pole when their coordinate extensiveness is primarily spacelike; for by the constancy of the critical velocity c, these data are unavailable for direct physical objectification. Data whose coordinate extensiveness is primarily timelike, on the other hand, are available for direct physical objectification, and are thus of a more public character. Similarly, in the mental pole there is a sense in which the world as an asymmetrically logically ordered global history is sheerly public, in that nonlocal, indirect prehensions are operative in the processes of decoherence and transmutation. But the privacy of data extensively coordinated as spacelike in the physical pole carries over into the mental pole at least in some sense; for the process of negative selection/negative prehension associated with decoherence and transmutation typically applies to the same data extensively coordinated as "spacelike-separated" and "private" in the physical pole.

Thus, in the same way that subject-relative "qualitative intensity" in the mental pole has its analogous, superject-relative "quantitative intensity" in the physical pole, one also finds that subject-relative privacy in the mental pole has its analogous superject-relative privacy in the physical pole; and superject-relative publicity in the physical pole has its analogous superject-relative publicity in the mental pole. This is, once again, a feature of the dipolarity of concrescence.

The theory of prehensions is founded upon the doctrine that there are no concrete facts which are merely public, or merely private. The distinction between publicity and privacy is a distinction of reason, and is not a distinction between mutually exclusive concrete facts. The sole concrete facts, in terms of which actualities can be analysed, are prehensions; and every prehension has its public side and its private side. Its public side is constituted by the complex datum prehended; and its private side is constituted by the subjective form through which a private quality is imposed on the public datum. The separations of perceptual fact from emotional fact; and of causal fact from emotional fact, and from perceptual fact; and of perceptual fact, emotional fact, and causal fact, from purposive fact; have constituted a complex of bifurcations, fatal to satisfactory cosmology. The facts of nature are the actualities; and the facts into which the actualities are divisible are their prehensions, with their public origins, their private forms, and their private aims. But the actualities are moments of passage into a novel stage of publicity; and the coordination of prehensions expresses the publicity of the world, so far as it can be considered in abstraction from private genesis. Prehensions have public careers, but they are born privately.[25]

Though the subject of concrescence enjoys its privacy during the process of integration, the potentia integrated—potentia associated with every actuality in the world—are historically public. And though this world as extensively coordinated data in the physical pole is similarly public from the perspective of the subject prehending these data, this publicity is restricted; it consists of manifold "private" regions by virtue of their extensive coordination and the associated critical velocity c—regions that lie outside the prehending subject's past light cone. Put more simply, just as actual occasions, in their character as subjects, enjoy their privacy during concrescence, so do nexūs of actual occasions in their characters as superjects enjoy their privacy after concrescence. Were this not the case, the actual world would never be extensively coordinated as diverse nexūs among a multiplicity of occasions, or as diverse corpuscular societies within an environmental nexus; there would instead be just one nexus—the world as sheer multiplicity, universally objectified with uniform intensity.

The extensive spatiotemporal coordination of the actual world into distinct nexūs and corpuscular societies within environmental nexūs has its obvious importance in quantum mechanics, as discussed earlier. Decoherence and transmutation cannot occur without actualities extensively coordinated as "environmental" to those societal nexūs extensively coordinated as "measured system" and "detector." Also, recall the operation of the Category of Subjective Harmony and the Category of Subjective Intensity in the mental pole—those Categoreal Obligations associated with the subjective aim of "balanced complexity." Balanced complexity, discernible in the genetic analysis of concrescence, entails notions of "contrasts" and "contrasts of contrasts" conditioned by a balance of conceptual reproduction and conceptual reversion. This complexity is grounded in contrasts of (i) conceptual reproductions fueled in part by the physical publicity of data, and (ii) conceptual reversions fueled in part by the privacy of both the prehending subject *and* the privacy of some of the extensively coordinated data (i.e., data outside the past light cone of the prehending subject). This complexity manifest in genetic analysis has its analog in extensive coordinate analysis. One example is the existence of complex adaptive systems in nature, inasmuch as their operations depend heavily on their extensive coordinations of their actual worlds. Conceptual Reproduction and Reversion are manifest in complex adaptive systems as "physical regularity" and "physical diversity," respectively—but also sometimes as "conceptual regularity" and "conceptual diversity" when the complex adaptive system under consideration is an abstract entity such as an economic system or government. But of course such abstract conceptual entities cannot be separated from the associated conscious physical entities that originated them, and so even conceptual regularity and diversity have their roots in physical regularity and diversity. And all of these have their ultimate roots in conceptual reproduction and conceptual reversion. But the extensive coordinations of complex adaptive systems do not entail merely the fundamental publicity of systems or their environments; it entails, rather, a balance of publicity and privacy. Complex adaptive systems both form their environments, and are formed by them. They are both privately distinct from, and publicly constitutive of their environments.

The concepts of publicity and privacy operative in concrescence are closely associated with the function of potentia in the physical and mental pole. Their quantum mechanical function in the latter was discussed extensively in earlier chapters. But their function in extensive coordinate division in the physical pole is equally significant. Since invariant spacetime intervals can be decomposed into relativistically noninvariant spacelike and timelike coordinations, the latter are *potential* spatial and temporal coordinations of *actual* invariant spacetime intervals whose actuality derives from the actuality of those entities related within these invariant intervals. Whitehead terms such potentia "eternal objects of the objective species." Because they derive from actualities and their invariant spacetime intervals, they are strictly relational of public data.

> Thus a member of this species can only function relationally: by a necessity of its nature it is introducing one actuality, or nexus, into the real internal constitution of another actual entity. Its sole avocation is to be an agent in objectification. It can never be an element in the definiteness of a subjective form. The solidarity of the world rests upon the incurable objectivity of this species of eternal objects.[26]

Potentia of the objective species do not, in other words, introduce creative reversions of the sort sometimes associated with potentia in the mental pole—those associated, for example, with quantum mechanical indeterminacy. These latter potentia are "eternal objects of the subjective species." Such a potential can function relatively, like potentia of the objective species; but they also function subjectively, as evinced by their quantum mechanical operations. A potential of the subjective species

> can be a private element in a subjective form, and also an agent in the objectification. In this latter character it may come under the operation of the Category of Transmutation and become a characteristic of a nexus as objectified for a percipient. . . . The intensity of physical energy belongs to the subjective species of eternal objects, but the peculiar form of the flux of energy belongs to the objective species.[27]

Potentia of the objective species are associated with the decomposition of invariant spacetime intervals into potential alternative spacelike and timelike coordinations which are relativistically noninvariant. These potentia might entail, as discussed previously, relativistic time dilations or length contractions. A potential of the objective species

> introduces into the immediate subject other actualities. The definiteness with which it invests the external world may, or may not, conform to the real internal constitutions of the actualities objecti-

fied. But conformably, or non-conformably, such is the character of that nexus for that actual entity. This is a real physical fact, with its physical consequences.[28]

The difficulty is, there is no way to explain the variances in the *qualitative intensities* of these coordinate objective potentia in the physical pole. For qualitative intensities pertain to subjective forms, whose valuations occur only in the mental pole—that is, the conceptual reproduction and reversion of subjective forms.

> The feelings—or more accurately, the quasi-feelings—introduced by the coordinate division of actual entities eliminate the proper status of the subjects entertaining the feelings. For the subjective forms of feelings are only explicable by the categoreal demands arising from the unity of the subject. Thus the coordinate division of an actual entity produces feelings whose subjective forms are partially eliminated and partially inexplicable. But this mode of division preserves undistorted the elements of definiteness introduced by eternal objects of the objective species.[29]

In other words, there must be some way in which to explain the relativistically noninvariant intensities or magnitudes of the potential spacelike and timelike decompositions in the physical pole, *prior to* the integration of these data into valuated, transmuted, quantum mechanical outcome states in the reduced density matrix. For these noninvariant coordinate intensities in the physical pole are always correlated with, and in some sense ingredient in, those valuated intensities in the mental pole. One method was to correlate qualitative intensity in the mental pole with quantitative intensity in the physical pole, with respect to the quantity of data extensively coordinated as "environmental" and the function of these data in the negative selection process associated with decoherence and transmutation. And the extensive spatiotemporal coordination of data as "environmental," as mentioned earlier, is predicated upon the constancy of the critical velocity c—a seemingly arbitrary physical requirement of the metaphysical scheme when compared with the deductively generated categoreal obligations; but it is a necessary physical requirement nonetheless:

> Thus in so far as the relationships of these feelings require an appeal to subjective forms for their explanation, the gap must be supplied by the introduction of arbitrary laws of nature regulating the relations of

intensities. Alternatively, the subjective forms become arbitrary epi-phenomenal facts, inoperative in physical nature, though claiming op-erative importance.[30]

Finally, since potentia of the objective species are always associ-ated with the coordinate decomposition of invariant spacetime inter-vals into relativistically noninvariant yet comparable, potential spacelike and timelike intervals, there must be a common morpho-logical structure underlying these intervals—a structure independent of the particular events constitutive of the intervals. As discussed earlier, this structure is a scheme of abstract extensive elements:

> The order of nature, prevalent in the cosmic epoch in question, exhib-its itself as a morphological scheme involving eternal objects of the objective species. The most fundamental elements in this scheme are those eternal objects in terms of which the general principles of coor-dinate division itself are expressed. These eternal objects express the theory of extension in its most general aspect. In this theory the no-tion of the atomicity of actual entities, each with its concrescent pri-vacy, has been entirely eliminated. We are left with the theory of extensive connection, of whole and part, of points, lines, and surfaces, and of straightness and flatness.[31]

In summary, coordinate division entails the decomposition of (i) the actual world as a manifold of invariant spacetime intervals among actualities into (ii) alternative, standpoint-dependent, poten-tial spacelike and timelike intervals—potential extensive coordina-tions of the data of the subject's actual world, relative to the standpoint of that subject. Coordinate division thus entails the prin-ciple of causal efficacy; for the alternative potential extensive coordi-nations effected by the subject are derived from the decomposition of an *asymmetrically, causally ordered manifold of invariant spacetime intervals*, and are therefore always governed by this asymmetrical order. Thus the order of causality is maintained among causally re-lated events, regardless of the relativistic noninvariance of the di-verse potential coordinations generated by the percipient subject.

> Thus in coordinate division we are analysing the complexity of the occasion in its function of an efficient cause. It is in this connection that the morphological scheme of extensiveness attains its impor-tance. In this way we obtain an analysis of the dative phase in terms of the "satisfactions" of the past world. These satisfactions are system-

atically disposed of in their relative status, according as *one* is or is not, in the actual world of *another*. Also they are divisible into prehensions which can be treated as quasi-actualities with the same morphological system of relative status. This morphological system gains special order from the defining characteristic of the present cosmic epoch. The extensive continuum is this specialized ordering of the concrete occasions and of the prehensions into which they are divisible.[32]

THE ABSTRACTIVE GEOMETRICAL SCHEME
UNDERLYING EXTENSIVE CONNECTION

Whitehead's main conceptual objection to general relativity was its derivation of extensive geometrical relations from actualities. It was discussed earlier, for example, that in general relativity, a "straight segment" is the straightest possible path between two spacetime events such that the segment forms a geodesic—i.e., the path whose timelike interval between the two events is of greatest magnitude when compared to any other interval connecting the two events. These intervals, in other words, involve the least amount of acceleration relative to any other interval connecting the events. Of this concept of measurement given in general relativity, Whitehead writes:

> The modern procedure, introduced by Einstein, is a generalization of the method of "least action." It consists in considering any continuous line between any two points in the spatio-temporal continuum and seeking to express the physical properties of the field as an integral along it. The measurements which are presupposed are the geometrical measurements constituting the coordinates of the various points involved. Various physical quantities enter as the "constants" involved in the algebraic functions concerned. These constants depend on the actual occasions which atomize the extensive continuum. The physical properties of the medium are expressed by various conditions satisfied by this integral.
>
> It is usual to term an "infinitesimal" element of this integral by the name of an element of distance. But this name, though satisfactory as a technical phraseology, is entirely misleading. There can be no theory of the congruence of different elements of the path. The notion of coincidence does not apply. There is no systematic theory possible, since the so-called "infinitesimal" distance depends on the actual

entities throughout the environment. The only way of expressing such so-called distance is to make use of the presupposed geometrical measurements. . . . The current physical theory presupposes a comparison of so-called lengths among segments without any theory as to the basis on which this comparison is to be made. . . . Further, the fact is neglected that there are no infinitesimals, and that a comparison of finite segments is thus required. [*Whitehead here refers to Einstein's direct association of gravity with the curvature of spacetime, such that gravitational acceleration is measured as a spacetime "distance."*] For this reason, it would be better—so far as explanation is concerned—to abandon the term "distance" for this integral, and to call it by some such name as "impetus," suggestive of its physical import.

It is to be noted, however, that the conclusions of this discussion involve no objection to the modern treatment of ultimate physical laws in the guise of a problem in differential geometry. The integral impetus *is* an extensive quantity, a "length." The differential element of impetus is the differential element of systematic length weighted with the individual peculiarities of its relevant environment. The whole theory of the physical field is the interweaving of the individual peculiarities of actual occasions upon the background of systematic geometry. This systematic geometry expresses the most general "substantial form" inherited throughout the vast cosmic society which constitutes the primary real potentiality conditioning concrescence.[33]

As discussed earlier, however, the general theory of relativity cannot be put into use without the imposition of absolute global boundary/symmetry conditions—that is, a preferred coordinate system of reference. And thus Whitehead's objection finds its relevance. The method by which Whitehead derives the extensive geometrical elements "point," "line," and "plane" among other such elements, *without* their derivation from actual occasions, is his "method of extensive abstraction," which begins with the notion of the possible "extensive connections" among and within "regions." An "abstractive set" is a set of regions where (i) any two members of the set are related such that one includes the other nontangentially (i.e., the way that a sheet of paper "includes" a box drawn on it); and (ii) there is no region included in *every* member of the set. Abstractive sets are, in other words, infinitely divisible regions; there is always some region within every region in an abstractive set. The regions are therefore serially ordered, much the same way that enduring objects and other societies of occasions are serially ordered. Thus the metaphysical scheme entails a close association between abstract extension and physical extension:

> It will be found that, though an abstractive set must start with *some* region at its big end, these initial large-sized regions never enter into our reasoning. Attention is always fixed on what relations occur when we have proceeded *far enough down* the series. The only relations which are interesting are those which, if they commence anywhere, continue throughout the remainder of the infinite series.[34]

An abstractive set α will "cover" abstractive set β when every member of α includes some members of β. And if α and β cover each other, then these abstractive sets are "equivalent." Finally, a "geometrical element" such as a point, line, or plane, is a complete group of abstractive sets that are mutually equivalent, and nonequivalent to any other abstractive set. A geometrical element such as a point can be "incident" in another geometrical element, such as a line. And a point is the geometrical element that has no other geometrical element incident within it. A "straight segment" is a linear geometrical element defined by a set of two points as endpoints of the segment. If the geometrical element contains three points, then it is a "flat segment" (i.e., a plane). Finally, a "complete locus" of points is a set of points that compose either all the points situated in a region, or the surface of a region, or incident in a geometrical element. Whitehead derives these and other abstract extensive geometrical elements in great detail in Part IV of *Process and Reality*, but the abbreviated presentation above should suffice for the following discussion.

For Whitehead, "extension" is "extensive connection." It is "a form of relationship between the actualities of a nexus. A point is a nexus of actual entities with a certain 'form'; and so is a 'segment.' Thus geometry is the investigation of the morphology of nexūs."[35] The associated notions of extensive connection and coordinate division in the physical pole have their connection to the association of quantum mechanics and genetic division in the mental pole; and this connection is analogous to the connection between "extensive connection" and "physical transmission." For Whitehead, physical transmission is the extensive connection of "contiguous" actual occasions:

> Let two actual occasions be termed "contiguous" when the regions constituting their "standpoints" are externally connected. Then by

reason of the absence of intermediate actual occasions, the objectification of the antecedent occasion in the latter occasion is peculiarly complete. There will be a set of antecedent, contiguous occasions objectified in any given occasion; and the abstraction which attends every objectification will merely be due to the necessary harmonizations of these objectifications. The objectifications of the more distant past will be termed "mediate"; the contiguous occasions will have "immediate objectification." The mediate objectifications will be transmitted through various routes of successive immediate objectifications. Thus the notion of continuous transmission in science must be replaced by the notion of immediate transmission through a route of successive quanta of extensiveness. These quanta of extensiveness are the basic regions of successive contiguous occasions.[36]

Thus quantum nonlocal effects of the sort evinced by EPR-type experiments are rare examples of direct or immediate objectification of one occasion by another when these occasions are physically, extensively noncontiguous. Most physical transmissions studied in physics, by contrast, entail immediate objectifications among contiguous extensive regions. Though generally restricted to the special arenas of quantum physics and the study of collisionless space plasmas,[37] quantum nonlocal phenomena are wholly compatible with Whiteheadian metaphysics:

> It is not necessary for the philosophy of organism entirely to deny that there is direct objectification of one occasion in a later occasion which is not contiguous to it. Indeed, the contrary opinion would seem the more natural for this doctrine. Provided that physical science maintains its denial of "action at a distance," the safer guess is that direct objectification is practically negligible except for contiguous occasions; but that this practical negligibility is a characteristic of the present cosmic epoch, without any metaphysical generality.[38]

A related concept with respect to quantum nonlocality is the distinction between the two species of physical prehensions associated with "physical transmission": "pure physical prehensions" and "hybrid physical prehensions":

> A pure physical prehension is a prehension whose datum is an antecedent occasion objectified in respect to one of its own *physical* prehensions. A hybrid prehension has as its datum an antecedent occasion objectified in respect to a *conceptual* prehension. Thus a pure

physical prehension is the transmission of a physical feeling, while hybrid prehension is the transmission of mental feeling.[39]

Nonlocal quantum mechanical interactions, whose analysis pertains to the operations of the mental pole—that is, conceptual reproductions, reversions, and transmutations—are examples of hybrid physical prehensions. By contrast, pure physical prehensions are typically characteristic of the non–quantum mechanical interactions described by classical mechanics, and of local quantum mechanical interactions—those which involve prehensions of contiguous antecedent quanta. These interactions are conditioned heavily by extensive coordinate division in the physical pole, where relativistic restrictions, such as the speed-of-light limitation, restrict causal influence of actualization by pure physical prehensions of antecedent actualities. And the nonlocal quantum interactions operative in the mental pole are primarily conditioned by the logical, historical, asymmetrical order of actualities yielded by genetic analysis. Put another way, if what is prehended in a contiguous antecedent occasion is the object occasion's own physical prehensions as operative in its own physical pole, then the prehension is pure. And if what is prehended in a contiguous or noncontiguous antecedent occasion is a conceptual prehension operative in the object occasion's own mental pole, then the prehension is hybrid.

> There is no reason to assimilate the conditions for hybrid prehensions to those for pure physical prehensions. Indeed the contrary hypothesis is the more natural. For the conceptual pole does not share in the coordinate divisibility of the physical pole, and the extensive continuum is derived from this coordinate divisibility. Thus the doctrine of immediate objectification for the mental poles and of mediate objectification for the physical poles seems most consonant to the philosophy of organism in its application to the present cosmic epoch.
>
> · But of course such immediate objectification is also reinforced, or weakened, by routes of mediate objectification. Also pure and hybrid prehensions are integrated and thus hopelessly intermixed. Hence it will only be in exceptional circumstances that an immediate hybrid prehension has sufficient vivid definition to receive a subjective form of clear conscious attention.[40]

But, as discussed earlier, the logical, historical, asymmetrical order operative in the mental pole has its reflection, too, in the physical pole, in the spatiotemporally coordinate invariant spacetime intervals *prior* to their decomposition/division into relativistically noninvariant spacelike and timelike coordinations. This relationship between the invariant spacetime intervals and their noninvariant, standpoint-dependent divisions is central to the relationship between "durations" and "strain-loci," respectively, which will be discussed presently.

The close relationship between extensive coordinate division as pertinent to the physical pole and quantum mechanical, genetic division as pertinent to the mental pole, and their associated notions of publicity and privacy, is summarized by Whitehead in the following passages:

> We have now traced the main characteristics of that real potentiality from which the first phase of a physical occasion takes its rise. These characteristics remain inwoven in the constitution of the subject throughout its adventure of self-formation. The actual entity is the product of the interplay of physical pole with mental pole. In this way potentiality passes into actuality, and extensive relations mould qualitative content and objectifications of other particulars into a coherent finite experience. . . .[41]

> It is by means of "extension" that the bonds between prehensions take on the dual aspect of internal relations, which are yet in a sense external relations. It is evident that if the solidarity of the physical world is to be relevant to the description of its individual actualities, it can only be by reason of the fundamental internality of the relationships in question. On the other hand, if the individual discreteness of the actualities is to have its weight, there must be an aspect in these relationships from which they can be conceived as external, that is, as bonds between divided things. The extensive scheme serves this double purpose.[42]

Apart from the extensive scheme, there would be no way to bridge (i) the solidarity of the world as characterized by the logical, historical ordering of actualities in the mental pole—that is, quantum mechanical characterizations of the world as a global history—with (ii) the relativistically *restricted* publicity of the world, given as causally influential relations among spatiotemporally divided actualities and nexūs of actualities, such that the causal agent always lies in the past light cone of the causal patient. Spatiotemporal privacy is guaranteed by this restricted publicity; but more generally, without the extensive scheme of relations, the privacy of identity enjoyed by all actualities—privacy of "individual discreteness"—would be entirely lost amid the sheer publicity of the logically, historically ordered world bound *only* by its quantum mechanical relations.

But the extensive relations exist, and despite the enticing illuminations quantum theoretical innovations have yielded in the physical sciences, these extensive relations cannot be discounted. Nature is not one universal, undifferentiated organism, whose distant connections are as intense as its local connections, in the sense that pain in one's hand is felt as immediately as pain in one's leg, or head. The universe is not alive, nor is it mind, nor is it disembodied information. The universe is a complex, extensive system, whose relations are both external, between actualities extensively and coordinately, and internal within actualities, intensively and genetically:

> The Cartesian subjectivism in its application to physical science became Newton's assumption of individually existent physical bodies, with merely external relationships. We diverge from Descartes by holding that what he has described as primary *attributes* of physical bodies are really the forms of internal relationships *between* actual occasions, and *within* actual occasions. Such a change of thought is the shift from materialism to organism, as the basic idea of physical science.
>
> In the language of physical science, the change from materialism to "organic realism"—as the new outlook may be termed—is the displacement of the notion of static stuff by the notion of fluent energy. Such energy has its structure of action and flow, and is inconceivable apart from such structure. It is also conditioned by "quantum" requirements. These are the reflections into physical science of the individual prehensions, and of the individual actual entities to which these prehensions belong. Mathematical physics translates the saying of Heraclitus, "All things flow," into its own language. It then becomes, All things are vectors. Mathematical physics also accepts the atomistic doctrine of Democritus. It translates it into the phrase, All flow of energy obeys "quantum" conditions.
>
> But what has vanished from the field of ultimate scientific conceptions is the notion of vacuous material existence with passive endurance, with primary individual attributes, and with accidental adventures. Some features of the physical world can be expressed in that way. But the concept is useless as an ultimate notion in science, and in cosmology.[43]

FACTS AND FORMS: DURATIONS AND LOCI

The application of Whitehead's theory of extensive connection to coordinate division, and its relation to the special and general theories of relativity discussed earlier in this chapter, entails two closely related sets of concepts. Each set contains an actual-factual term, paired with a potential-formal term: (i) a "duration" of actual facts, and its potential, formal "complete locus"; and (ii) a "presented duration" of actual facts, and its potential, formal "strain-locus."

A "duration" is a nexus of contemporary actualities in "unison of becoming." The duration is a nexus of facts that entails certain formal extensive, geometrical relations, and the whole of these formal relations is referred to as a "complete locus." If the universe described in relativity theory comprises manifold relativistically invariant spacetime intervals, each comprising logically, historically, asymmetrically ordered actualities, then a duration is a cross section of the actualities comprised. But because there can be multiple invariant spacetime intervals that include a particular actual occasion (consider the worldlines of each twin sister intersecting at the departure event of the spacefaring sister), it is possible that different occasions might be contemporaries of an event M, but might not be contemporaries of each other. In other words, M can lie in more than one duration. This is clearly reflected in the decomposition of invariant spacetime intervals into relativistically noninvariant spacelike and timelike intervals; for example, events M and N might lie outside the past and future light cone of event E, being contemporaries of E; but suppose also that M lies *inside* the past light cone of N. In such a case, M and N are both contemporaries of E, but not contemporaries of each other, since M is in the causal past of N. Were it not for relativity, then, each actual occasion would lie in one and only one duration.

A duration is a complete locus of actual occasions in "unison of becoming," or in "concrescent unison." It is the old-fashioned "present state of the world." In reference to a given duration, D, the actual world is divided into three mutually exclusive loci. One of these loci is the duration D itself. Another of these loci is composed of actual occasions which lie in the past of some members of D: this locus is the "past of the duration D." The remaining locus is composed of actual occasions which lie in the future of some members of D: this locus is the "future of the duration D."

By its definition, a duration which contains an occasion M must lie within the locus of the contemporaries of M. According to the classical pre-relativistic notions of time, there would be only one duration including M, and it would contain all M's contemporaries. According to modern relativistic views, we must admit that there are many durations including M—in fact, an infinite number, so that no one of them contains all M's contemporaries.

Thus the past of a duration D includes the whole past of any actual occasion belonging to D, such as M for example, and it also includes *some* of M's contemporaries. Also the future of the duration D includes the whole future of M, and also includes *some* of M's contemporaries.

So far, starting from an actual occasion M, we find six loci, or types of loci, defined purely in terms of notions derived from "causal efficacy." These loci are, M's causal past, M's causal future, M's contemporaries, the set of durations defined by M; and finally, taking any one such duration which we call D as typical, there is D's past, and D's future. Thus there are the three definite loci, the causal past, the causal future, and the contemporaries, which are defined uniquely by M; and there are the set of durations defined by M, and the set of "durational pasts" and the set of "durational futures." The paradox which has been introduced by the modern theory of relativity is twofold. First, the actual occasion M does not, as a general characteristic of all actual occasions, define a unique duration; and secondly, such a unique duration, if defined, does not include all the contemporaries of M.[44]

Each duration is associated with a "complete locus" of actualities, in the sense that it is a cross section of all invariant spacetime intervals constituting the universe. The invariance of spacetime intervals, then, allows for the asymmetrical, logical ordering of the durations intersecting them; thus, asymmetrically and logically associated with each duration is a set of "durational pasts" and "durational futures."

"Presented durations," however, are relativistically noninvariant, propositionally restricted durations—that is, restricted to the frame of reference of a particular actuality. In contrast with a duration, which consists of a *complete locus* of actualities in concrescent unison, a presented duration is intimately related with (but not equivalent to) a *propositional locus* or "strain-locus" of *potential relations* among *potential contemporaries*. Presented durations and their associated potentia, like the potentia involved in quantum mechanical state evolution, are operative in the *mental pole*, where they are conditioning ingredients in the processes of integration, reintegration, valuation, and transmutation associated with that pole. But their subjective formal temporal valuations in the mental pole as "presented durations," and spatial valuations as "strain-loci," are conditioned by the coordinate division, in the physical pole, of invariant spacetime intervals into potential, relativistically noninvariant, potential timelike intervals and spacelike intervals.

These relativistically noninvariant, standpoint-dependent intervals are *potential intervals* and thus operative in the mental pole; but they originated, in part, from the coordinate division operative in the physical pole—that is, the decomposition into potential spacelike and timelike intervals, governed by and, in a sense, *valuated by* certain laws of physical nature. The constancy of the critical velocity *c* is intimately bound up with these laws. Such valuation is typically related to subjective forms in the mental pole, as exemplified quantum mechanically in previous chapters; but the conditioned spatiotemporal potentia referred to here occur via coordinate division in the physical pole, and therefore require an explanation. Again, the appeal is to the operation of physical laws in nature:

> Thus in so far as the relationships of these feelings require an appeal to subjective forms for their explanation, the gap must be supplied by the introduction of arbitrary laws of nature regulating the relations of intensities. Alternatively, the subjective forms become arbitrary epiphenomenal facts, inoperative in physical nature, though claiming operative importance.[45]

Indeed, the close association of valuation by physical laws in the physical pole and the quantum mechanical probability valuations in the mental pole is clearly evinced by the fact that even nonlocal quantum mechanical valuations are compatible with special relativity. Relativistic quantum field theory is one example of this compatibility.

A presented duration, then, is the product of coordinate decomposition, in the physical pole, of the invariant spacetime intervals constitutive of the universe into standpoint-dependent, *timelike* coordinations of actual facts. The temporality associated with a duration as a set of "mutual contemporaries" requires that the duration have "factual" or "physical" content; for in Whiteheadian metaphysics, "temporality" is derived from serial historical routes of *actualities*.

> A duration is a complete set of actual occasions, such that all the members are mutually contemporary one with the other. This property is expressed by the statement that members enjoy "unison of immediacy." . . . Every occasion outside the set is in the past or in the future of some members of the set, and is contemporary with other members of the set. According as an occasion is in the past, or the future, of some members of a duration, the occasion is said to be in the past, or in the future, of that duration.
>
> No occasion can be both in the past and in the future of a duration. Thus a duration forms a barrier in the world between its past and its future.[46]

Durations are coordinately decomposed or divided into *temporal*, standpoint-dependent, presented durations. And in the same way, the "complete locus" of actualities associated with durations is decomposed into a standpoint-dependent, *spatial* coordination, or "strain-locus." A "strain" is simply the prehension of actualities, and their relations, as *geometrically coordinated*—that is, as decomposed from invariant spacetime intervals into standpoint-dependent spacelike intervals.

Whereas a duration entails *physical factual content*, a strain-locus only entails *geometrical formal content*. A "strain-locus" is a set of mutually exclusive and exhaustive straight-line "projectors" that join all the pairs of points in the local, spatially coordinated region. The projectors are merely the spacelike intervals yielded by the decomposition of invariant spacetime intervals into (i) a *presented duration*, or relativistically noninvariant, standpoint-dependent *temporal coordination*; and (ii) a *strain-locus*, or relativistically noninvariant, standpoint-dependent *spatial coordination*. That the abstract geometrical elements of strain-loci are independent of the physical facts they coordinate reflects the importance in Whitehead's metaphysics that geometrical elements such as points, lines, and planes, are a priori abstract extensive elements. They are related to the requirement of universal boundary or symmetry conditions in general relativity, in that they provide a uniform scheme of reference by which congruence relations among diverse spacetime intervals can be made. A strain-locus is

> a systematic whole, independently of the actualities which may atomize it. In this it is to be distinguished from a "duration" which does depend on its *physical content*. A strain-locus depends merely upon its *geometrical content*. This geometrical content is expressed by any adequate set of "axioms" from which the systematic interconnections of its included straight lines and points can be deduced. This conclusion requires the systematic uniformity of the geometry of a strain-locus, but refers to further empirical observation for the discovery of the particular character of this uniform system. . . .[47]

> This locus is the reason why there is a certain absoluteness in the notions of rest, velocity, and acceleration. For this presented duration is the spatialized world in which the physical object is at rest, at least momentarily for its occasion M.[48]

One may, then, correlate (i) the general relativistic decomposition of an invariant spacetime interval into noninvariant (subject-dependent/propositional) timelike and spacelike intervals, and (ii) Whiteheadian coordinate division of concrescence into presented duration and strain-locus, respectively. The strain-locus is a propositional set of spacelike geometrical intervals (projectors) and these are subject dependent (i.e., relativistically noninvariant). They derive not from the "actual" content of the world, but rather from the potential forms that a subject might project upon its immediate world. A presented duration is, by the same analogy, a propositional timelike feature as regards general relativity. A presented duration pertains to that frame of reference wherein the prehending subject is "at rest" relative to its actual world; and in general relativity, the concept of the "proper time" of a spacetime interval is defined in exactly this way. Thus, the "straightest possible worldline" describing a serial route of two occasions is a geodesic that spans both occasions where each is at rest relative to the other. Because there is no relative motion or acceleration associated with the two occasions, the timelike interval spanning the two occasions will be maximal (i.e., there is no time dilation). In general relativity, such an interval forms the "straightest possible" worldline, and in this way, geometry is deeply connected to temporality; space and time are not complementary qualifications of actuality, but are mutually implicative features of spacetime. Presented durations and strain-loci are, for Whitehead, similarly mutually implicative.

Consider, again, the notion that every actual occasion is, in some sense, always "at rest" relative to itself. Such a notion of "at rest," if it is to have any real meaning, requires relativity not just between an occasion and its history as restricted to itself, but between an occasion and its world. Any meaningful notion of "at rest" or "presented duration," in other words, implies spatial, geometrical relations between the subject and its actual world. Thus "presented duration" and "strain-loci" are mutually implicative in the same way that "proper time" and geodesic "straightness" are mutually implicative in general relativity.

As regards both Whiteheadian spatiotemporal extensiveness and relativity theory, it is always potential forms that are both decomposed and integrated, not actual facts—just as is the case with quantum mechanics as discussed in previous chapters. Potentia, in other words, are as ontologically significant as actualities in both quantum mechanics and relativity theory—a fact which is crucial to their accommodation by Whiteheadian metaphysics. Recall, for example, that the decompositions described in figures 5.1 and 5.2 are propositional:

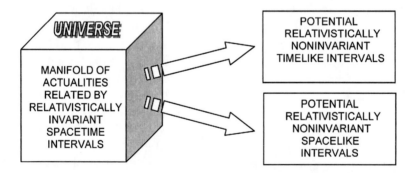

FIGURE 5.1. Relativistic coordinate division in the physical pole. The standpoint-independent spacetime world is coordinately decomposed/divided into a standpoint-dependent world of actualities with potential relations coordinated as spacelike and timelike. There are two analogous presuppositions underlying such division: (i) a priori universal symmetry or boundary conditions in the general theory of relativity, and (ii) a universal scheme of abstract extension in Whiteheadian metaphysics.

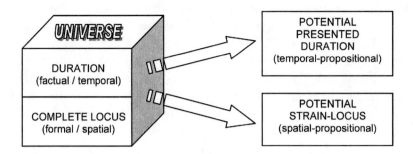

FIGURE 5.2. Reflection of coordinate division into the mental pole. The coordinate decomposition of the standpoint-independent spacetime world in the physical pole (fig. 5.1) yields a universe initially coordinated in the mental pole as (i) a duration with factual content, related to (ii) a complete locus of formal, spatial relations among the duration's actualities. (The duration is "factual" and "temporal" because in Whiteheadian metaphysics, "temporality" is derived from serial historical routes of actualities.) The world as duration/complete locus evolves in the mental pole to become a standpoint-dependent world of actualities with relations potentially coordinated as timelike (via the presented duration) and spacelike (via the strain-locus).

> When we divide the satisfaction coordinately, we do not find feelings which *are* separate, but feelings which *might be* separate. In the same way, the divisions of the region are not divisions which *are*; they are divisions which *might be*. Each such mode of division of the extensive region yields "extensive quanta": also an "extensive quantum" has been termed a "standpoint."[49]

There is, then, a close relationship between these strain-locus projectors and the quantum mechanical projectors (those described in chapters 2 and 3); they are cooperative in the mental pole. Both types of projector constitute *potential forms of facts*. The coordinate divisions of spacetime are propositional in that they yield *standpoint-dependent*, relativistically noninvariant coordinate spacelike and timelike intervals. These are reflected in the mental pole by strain-loci and presented durations, respectively—potential forms of facts that become integrated and valuated as propositional transmutations of data. These transmutations are cooperative with the quantum mechanical transmutations of potentia into unified potential quantum histories valuated as probabilities. Thus the valuated potential subjective forms or transmuted propositional *unifications* of data in the mental pole are correlated with relativistic, propositional coordinate *divisions* in the physical pole. These potential divisions, reflected in the presented durations and strain-loci of the mental pole, are as ontologically significant as the potential outcome states yielded in quantum mechanics; they are not merely epistemically significant, or worse, potential "distortions" of the objective, actual world:

> This spatialization is a real factor in the physical constitution of every actual occasion belonging to the life-history of an enduring physical object. . . . The reality of the rest and the motion of enduring physical objects depends on this spatialization for occasions in their historic routes. The presented duration is the duration in respect to which the enduring object is momentarily at rest. It is that duration which is the strain-locus of that occasion in the life-history of the enduring object.[50]

The spatially and temporally extensive potentia associated with strain-loci and presented durations introduce

> into the immediate subject other actualities. The definiteness with which it invests the external world may, or may not, conform to the real internal constitutions of the actualities objectified. But conformably, or non-conformably, such is the character of that nexus for that actual entity. This is a real physical fact, with its physical consequences.[51]

This ontologically significant, cooperative association of strain-locus projectors and quantum mechanical projectors is exemplified, for example, in the phenomenon of quantum tunneling, central to devices such as the scanning tunneling microscope (STM) and experimental quantum transistors. Quantum tunneling occurs when a particle such as an electron, spatiotemporally coordinated relative to a spatiotemporally coordinated barrier (such as an energy well associated with the nuclear forces within an atom), actualizes across the barrier rather than having overcome it with the requisite energy required to do so. Thus, the quantum state change of the system entails a quantum change in the spatiotemporal coordination of the actualities comprised by the system. The spatiotemporal coordinations of these actualities, in other words, do not endure through a quantum measurement interaction (concrescence); rather, they evolve with it. Thus the probability valuations associated with the potential quantum mechanical outcome states of the system—including the probability that the electron will actualize across the barrier—are intimately bound up with the valuated spatiotemporal potentia—the alternative potential strain-loci, for example. Again, these potentia are operative in the mental pole, but have their origins in extensive coordinate division, operative in the physical pole.

Thus, conceptual reproductions, reversions, and transmutations pertain to strain-loci as well as to the potentia associated with quantum mechanics; and they have the same implications for the subjective aim of "balanced complexity" of the sort exemplified by complex adaptive systems in nature as discussed previously:

> In the process of integration, these wider geometrical elements acquire implication with the qualities originated in the simpler stages. The process is an example of the Category of Transmutation; and is to be explained by the intervention of intermediate conceptual feelings. . . .[52]

> But the growth of ordered physical complexity is dependent on the growth of ordered relationships among strains. Fundamental equations in mathematical physics, such as Maxwell's electromagnetic equations, are expressions of the ordering of strains throughout the physical universe. . . .[53]

> It is the mark of a high-grade organism to eliminate, by negative prehension, the irrelevant accidents in its environment, and to elicit massive attention to every variety of systematic order. For this purpose, the Category of Transmutation is the master-principle.[54]

It must be emphasized that strain-loci (potential geometrical forms of facts) and presented durations (the potential facts themselves), though they derive from the decomposition of invariant spacetime durations and complete loci, are not necessarily closely correlated. As is the case in quantum mechanics, potentia can be integrated in any number of ways even after decoherence and transmutation; one is always left with a set of valuated potential outcome states that entail different integrations of potentia.

> There is no reason, derivable from these definitions [of strain-loci and durations], why there should be any close association between the strain locus of an experient occasion and any duration including that occasion among its members. It is an empirical fact that mankind

invariably conceives the presented world as consisting of such a duration. This is the contemporary world as immediately perceived by the
senses. But close association does not necessarily involve unqualified
identification.[55]

Again, it is only by means of an a priori scheme of abstract geometrical relations that these diverse potential spatiotemporal integrations—diverse presented durations and strain-loci—are capable of
relation at all. Without such a scheme, there would be no congruence relations by which to correlate diverse spatiotemporal coordinations. Whitehead defines "congruence" thus: "Two segments are
congruent when there is a certain analogy between their functions in
a systematic pattern of straight lines, which includes both of
them."[56] He continues, "The only point to be remembered is that
each system of 'coordinates' must have its definable relation to the
analogy which constitutes congruence.'"[57]

The necessary presupposition of a scheme of abstract geometrical
relations in the general theory of relativity is evinced, as mentioned
earlier, in the requirement of an arbitrarily selected set of boundary
or symmetry conditions that must be applied as a background system
of reference for any measurement made by the theory. Indeed, were
it not for such a background, it is unlikely the universe would be
capable of any kind of rational coordination at all.

> It is by reason of this disclosure of ultimate system that an intellectual
> comprehension of the physical universe is possible. There is a system-
> atic framework permeating all relevant fact. By reference to this frame-
> work the variant, various, vagrant, evanescent details of the abundant
> world can have their mutual relations exhibited by their correlation to
> the common terms of a universal system. . . . For example, the conjec-
> ture by an eminent astronomer, based on measurements of photo-
> graphic plates, that the period of the revolution of our galaxy of stars
> is about three hundred million years can only derive its meaning from
> the systematic geometrical relations which permeate the epoch. But
> he would have required the same reference to system, if he had made
> an analogous statement about the period of revolution of a child's top.
> Also the two periods are comparable in terms of the system. . . .[58]

> . . . All exact measurements concern perceptions in the mode of pres-
> entational immediacy [and] . . . such observations purely concern the
> systematic geometric forms of the environment, forms defined by pro-
> jectors . . . irrespective of the actualities which constitute the environ-

ment. The contemporary actualities of the world are irrelevant to these observations. All scientific measurements merely concern the systematic real potentiality out of which these actualities arise. This is the meaning of the doctrine that physical science is solely concerned with the mathematical relations of the world.

These mathematical relations belong to the systematic order of extensiveness which characterizes the cosmic epoch in which we live. The societies of enduring objects—electrons, protons, molecules, material bodies—at once sustain that order and arise out of it. The mathematical relations involved in presentational immediacy thus belong equally to the world perceived and to the nature of the percipient. They are, at the same time, public fact and private experience.[59]

In other words, the spatiotemporal coordinations given as presented durations and strain-loci—coordinations that take place in the mental pole in the mode of "presentational immediacy"—have their roots in the historical routes of invariant spacetime intervals constitutive of the actual world coordinately divided in the physical pole. As such, these spatiotemporal potentia are often inherited directly from the antecedent world as "real potentia." Thus the possibility of rampant reversions—presented durations and strain-loci which have little or no relation whatsoever to the invariant spacetime intervals from which they derive—is balanced by the category of conceptual reproduction, driven by "real potential" geometrical coordinations inherited directly from past occasions. In Whiteheadian metaphysics, then, physical "inertia" derives from spatiotemporal conceptual reproductions; and physical "acceleration" derives from spatiotemporal conceptual reversions.

The indifference of presentational immediacy to contemporary actualities in the environment cannot be exaggerated. It is only by reason of the fortunate dependence of the experient and of these contemporary actualities on a common past, that presentational immediacy is more than a barren aesthetic display. It does display something, namely, the real extensiveness of the contemporary world.[60]

Notes

1. Robert J. Griffiths, *Stat. Phys.* 36 (1984): 219. See also Griffiths, *Consistent Quantum Theory* (Cambridge: Cambridge University Press, 2002),

and Roland Omnès, *The Interpretation of Quantum Mechanics* (Princeton, N.J.: Princeton University Press, 1994).

2. See, for example, B. Roy Frieden, *Physics from Fisher Information* (Cambridge: Cambridge University Press, 1999).

3. H. A. Lorentz, A. Einstein, H. Minkowski, and H. Weyl, *The Principle of Relativity: A Collection of Original Memoirs on the Special and General Theory of Relativity* (New York: Dover, 1952), 75.

4. Albert Einstein, "Zur Elektrodynamik bewegter Korper," *Annalen der Physik* 17 (1905): 892; trans. Arthur I. Miller, "On the Electrodynamics of Moving Bodies," appendix to *Albert Einstein's Special Theory of Relativity* (Reading, Mass.: Addison-Wesley, 1981), 393.

5. Alfred North Whitehead, *Process and Reality: An Essay in Cosmology, Corrected Edition*, ed. D. Griffin and D. Sherburne (New York: Free Press, 1978), 3–4.

6. Alfred North Whitehead, *The Concept of Nature* (Cambridge: Cambridge University Press, 1920), 193.

7. Ibid., 195.

8. H. Wildon Carr, R. A. Sampson, and A. N. Whitehead, "Relativity, Logic, and Mysticism," *Aristotelian Society, Supplementary* 3 (1923): 34–41.

9. Whitehead, *Process and Reality*, 302.

10. T. L. Heath, *Euclid in Greek* (Cambridge: Cambridge University Press, 1920), quoted in Whitehead, *Process and Reality*, 302.

11. Whitehead, *Process and Reality*, 303.

12. Ibid., 283.

13. Ibid., 288.

14. Ibid., 283–284.

15. Ibid., 284.

16. Ibid., 285.

17. Ibid., 284.

18. Ibid., 285.

19. Ibid., 285–286.

20. Ibid., 286.

21. Ibid., 288.

22. Ibid., 287.

23. Ibid., 289.

24. Ibid., 289.

25. Ibid., 290.

26. Ibid., 291.

27. Ibid., 291–292.

28. Ibid., 291.

29. Ibid., 292.

30. Ibid., 292.

31. Ibid., 292.
32. Ibid., 293.
33. Ibid., 332–333.
34. Ibid., 298.
35. Ibid., 302.
36. Ibid., 307.
37. Timothy E. Eastman, "Micro- to Macroscale Perspectives on Space Plasmas," *Phys. Fluids B* 5, no. 7: 2671–2675.
38. Whitehead, *Process and Reality*, 307–308.
39. Ibid., 307–308.
40. Ibid., 308.
41. Ibid., 308.
42. Ibid., 309.
43. Ibid., 309.
44. Ibid., 320.
45. Ibid., 292.
46. Ibid., 322.
47. Ibid., 330.
48. Ibid., 321.
49. Ibid., 283–284.
50. Ibid., 321.
51. Ibid., 291.
52. Ibid., 310.
53. Ibid., 311.
54. Ibid., 317.
55. Ibid., 323.
56. Ibid., 331.
57. Ibid., 332.
58. Ibid., 327.
59. Ibid., 326.
60. Ibid., 324.

6

Summary and Outlook

ANY BRIDGE intended to span the chasm separating classical and quantum mechanics must be constructed upon a sound ontological framework that is (i) coherent, in that its most fundamental concepts are incapable of abstraction from each other and thus free from self-contradiction; (ii) logical; (iii) empirically applicable; and (iv) empirically adequate, in that the ontology is applicable universally—both to the realm of familiar experience as well as to that of theoretical experience.

These are the desiderata Whitehead proposes for the ideal of speculative philosophy, of which his metaphysical scheme is an example. Critical evaluation of this scheme over the many years since it was introduced reveals that the first two requirements, the rational requirements that the scheme be coherent and logical, are satisfied; this is to be expected, as Whitehead constructed his philosophy with these requirements foremost in mind. And a successful correlation of his cosmological scheme with quantum mechanics contributes significantly to the satisfaction of the empirical requirements of applicability and adequacy. Again, this is to be expected, for it is clear that Whitehead was familiar with the innovations of quantum theory—at the very least, the quantum theory of Planck and Einstein and its relationship to Bohr's 1913 model of the atom—as he makes explicit reference to these and suggests specific correlations, including equivalences in terminology, between these theories and his metaphysical scheme.

The question as to whether this applicability and adequacy was similarly designed to accommodate the "new" quantum theory of Heisenberg, Schrödinger, Bohr, Born, Dirac, et al., though interesting and arguable, ultimately has little to do with whether or not the applicability and adequacy of Whitehead's cosmological scheme does in fact accommodate this "new" quantum theory, hashed out at the Solvay Conferences of 1927 and 1930 and in the years since. In that the "old" quantum theory is subsumed and better systematized by the new, the applicability of Whitehead's scheme to the accommodation of the latter is, at the very least, reasonable. Also, Whitehead explicitly refers to the mathematical terms and concepts used by the "new" quantum theory (though not exclusively, of course)—"the theory of probability, the theory of tensors, the theory of matrices."[1] He uses, for example, the term "matrix" to describe a set of alternative "subjective forms"[2] or qualified, valuated, mutually exclusive, potential integrations of facts; and this is a perfect analog of the reduced density matrix in quantum mechanics—a set of alternative "outcome states" or qualified, probability-valuated, mutually exclusive, potential integrations of facts.[3] These observations aside, Whitehead states many times in *Process and Reality* that his metaphysical scheme is empirically applicable and adequate to the task of accommodating modern physics, both applied (i.e., meeting the desideratum of empirical applicability) *and* theoretical (i.e., meeting the desideratum of empirical adequacy)—the physics of the time, and the physics of the foreseeable future. "The general principles of physics," Whitehead writes, "are exactly what we should expect as a specific exemplification of the metaphysics required by the philosophy of organism. . . . In this way, the philosophy of organism—as it should—appeals to the facts."[4]

The degree of success to which Whitehead's cosmological scheme is able to accommodate quantum mechanics coherently, logically, applicably, and adequately is best estimated by its ability to bridge quantum and classical conceptions of nature; for each of these conceptions by itself is valid in its own application to selected realms of experience. But if it is to be believed that either or both of these conceptions exemplify ontological first principles, then these principles must apply to both realms of experience in a coherent way, which by most interpretations they do not. The incompatibilities can be distilled, as suggested by Abner Shimony, into the following five conceptual innovations implied by quantum mechanics: (i) objective indefiniteness, (ii) objective chance, (iii) objective nonepistemic probability, (iv) objective entanglement, and (v) quantum nonlocality.[5]

Attempts at a coherent interpretation of these innovations have typically entailed one of the following two strategies:

1. A classical ontological interpretation of quantum mechanics, such that these five innovations can be dismissed as errors stemming

from the incompleteness of the quantum theory—errors remedied by a classical interpretation. This is the strategy evinced by any interpretation of quantum mechanics that characterizes indefiniteness, chance, entanglement, probability, and nonlocality as epistemic artifacts of quantum theory's inability to account for a multiplicity of "hidden variables." These variables, if disclosed, would complete the theory, thus cleansing it of all uncertainty so that its true essence as a fully determinate, classical theory would be revealed. Since by such an interpretation human limitations prevent the specification of these deterministic hidden variables, however, the incompleteness of the theory with its unmovable veil of indeterminism relegates us to a statistical approximation of them. Thus the matrix of probable outcome states yielded by quantum mechanics is to be interpreted, as advocated early on by Max Born, as statistical probabilities of purely epistemic significance[6]—probabilities that describe the statistical frequency at which an experimenter measuring the classical position of an electron, for example, will find it in a given region around an atomic nucleus after repeated experiments. It describes, in other words, the probability that one will find that the *pre-established facts* will fit a *probable form* a certain percentage of the time when a given experiment is repeated sufficiently.

This statistical or instrumental interpretation of quantum mechanics is thus able to account for indefiniteness, change, probability, and entanglement by rendering these concepts merely epistemically significant, as opposed to ontologically significant (as implied by Shimony's use of the adjective "objective" in his list).

2. Restriction of the characterization of the ontological significance of quantum mechanics, and classical mechanics, to the following principle: "Objective" reality is necessarily veiled, and the veil can be lifted only so high as to reveal the subjective coordinations of our experiences of this underlying, necessarily hidden reality. The apparent objectivity of physical laws derives from occasional widespread and thus "public" agreements with respect to these subjective coordinations; but since such objectivity is in truth merely a shared epistemic paradigm, it is to be expected that other "public" agreements might arise which prove to be incompatible. From this principle follows Bohr's Principle of Complementarity, which holds the conceptual incompatibilities separating the *fundamental* characterization of nature described by quantum and classical mechanics as

evidence of this sole ontological principle. There is thus no need for an ontological interpretation capable of accommodating both quantum and classical mechanics, even if one were possible; for each is but a public means of coordinating experiences in its exclusive realm, and there is no reason to suppose that these two complementary publics will ever meet.

Each of these strategies exemplifies what Whitehead describes as the pitfalls of applying the method of "philosophic generalization" to the natural sciences, which tends to end with such generalization rather than begin with it:

> The term "philosophic generalization" has meant "the utilization of specific notions, applying to a restricted group of facts, for the divination of the generic notions which apply to all facts." . . . In its use of this method natural science has shown a curious mixture of rationalism and irrationalism. Its prevalent tone of thought has been ardently rationalistic within its own borders, and dogmatically irrational beyond those borders. In practice such an attitude tends to become a dogmatic denial that there are any factors in the world not fully expressible in terms of its own primary notions devoid of further generalization. Such a denial is the self-denial of thought.[7]

In contrast with the two strategies above, however, there have been a number of proposed interpretations of quantum mechanics which have attempted to both embrace the validity of the principles exemplified by classical mechanics while at the same time proposing a more fundamental characterization of these principles and their application to the natural world. The value of classical mechanics is thus maintained by virtue of its abstraction from quantum mechanics, so that if the value of the latter is demonstrated, then the value of the former is secured as well, even though it is no longer fundamental.

These interpretations—many of which have been surveyed briefly in this book—apart from their various strengths and weaknesses, all exemplify the spirit of Whitehead's notion of "imaginative rationalization," of which "philosophic generalization" is merely the first step. For when it is the last step, as is the case with the various "classical" interpretations of quantum mechanics, the interpretation is confined strictly to the terms and concepts which preceded it. Important observations that typically rely on sharp contrasts, are, then, doomed to limitation:

Thus, for the discovery of metaphysics, the method of pinning down thought to the strict systematization of detailed discrimination, already effected by antecedent observation, breaks down. This collapse of the method of rigid empiricism is not confined to metaphysics. It occurs whenever we seek the larger generalities. In natural science this rigid method is the Baconian method of induction, a method which, if consistently pursued, would have left science where it found it. What Bacon omitted was the play of a free imagination, controlled by the requirements of coherence and logic. . . . The reason for the success of this method of imaginative rationalization is that, when the method of difference fails, factors which are constantly present may yet be observed under the influence of imaginative thought. Such thought supplies the differences which the direct observation lacks.[8]

Of those "imaginatively rationalized" interpretations of quantum mechanics that attempt to bridge the classical and quantum worlds in terms of the latter as fundamental—for example, the Relative State Interpretation of Everett; the Spontaneous Localization Theory of Ghirardi, Rimini, Weber, and Pearle; and the decoherence-based theories of Żurek, Gell-Mann, Omnès, and others—this discussion has, I hope, demonstrated that it is the decoherence-based theories that are the most coherent and logical of these imaginative rationalizations. Their superiority in terms of empirical applicability and adequacy has been, and will continue to be, the subject of a great deal of debate and experimentation.

But the correlation of Whitehead's metaphysical scheme and the decoherence-based interpretations of quantum mechanics only begins with these common desiderata. Careful analysis reveals that this family of interpretations of quantum mechanics can be characterized as a detailed and fundamental exemplification of Whitehead's philosophy of organism as it pertains to the physical sciences. Every phase of concrescence has its conceptual and mechanical analog in quantum mechanics; every Categoreal Obligation is presupposed and exemplified.

In Part I it was demonstrated how the decoherence-based interpretations reconcile Shimony's five conceptual quantum theoretical innovations with classical mechanics; this reconciliation, as we have seen, relies heavily upon the decoherence effect—particularly, (i) its processes of negative selection and elimination of interfering potentia via their environmental correlations, and (ii) its subsequent process integrating the remaining potential facts into mutually exclusive, probability-valuated forms of definiteness relative to a given subject/indexical eventuality. Enduring, material objects and their qualifications are thus more fundamentally described as historical routes of actualities (Whitehead's "social nexūs with personal order"—i.e., serial order) wherein processes of decoherence produce a high degree of reproduction and regularity in their highly valuated potentia. The "quasi-classicality" or apparent classicality of nature is, by these interpretations, predicated upon decoherence and its related processes.

The apparent classicality of nature is similarly accounted for by Whitehead's categoreal scheme, such that: (i) objective indefiniteness and (ii) objective chance each exemplifies and presupposes both Category V, *Conceptual Reversion*, and Category IX, *Freedom and Determination*; (iii) objective nonepistemic probability exemplifies and presupposes Category VII, *Subjective Harmony*, and Category VIII, *Subjective Intensity*; (iv) objective entanglement and (v) quantum nonlocality each exemplifies and presupposes Category I, *Subjective Unity* and Category VI, *Transmutation*, as well as the Principle of Relativity and the Ontological Principle.

The correlation of Whitehead's metaphysical scheme and the decoherence-based interpretations of quantum mechanics—particularly those that include the concept of "historical routes" of system states, analogous with Whitehead's historical routes of social nexūs (or "societies with personal order")—is particularly noteworthy with respect to the fifth of Shimony's conceptual innovations: quantum nonlocality. For as discussed in Part I, quantum mechanics allows for an actuality in Region A to affect the potentia associated with an actualization, or historical route of actualizations, constituting a system in some Region B. According to both quantum mechanics and Whiteheadian metaphysics, any fact in A that is logically prior to a concrescing actualization in B affects the integrations of potentia operative in B during its concrescence, even if A and B are spatially well separated.

But by this scheme, the nonlocal affection upon the concrescing fact in B by the antecedent fact in A is *not* manifest as a causal influence, and thus the spacetime structure of special relativity, incorporated by Whitehead into his cosmology, is not violated. This difference between the terms "nonlocal causal affection of potentia by logically prior actuality" and "local causal influence of actualization by temporally prior actuality" is acute, and often overlooked when interpreting quantum nonlocality. "Causal influence," in the Whiteheadian scheme, is operative in the physical pole or primary stage (the conformal phase, or phase of causal efficacy), and is bound by the speed of light according to the theory of special relativity; "causal affection" is operative in the mental pole or supplementary stage, and is not limited by special relativity. Causal influence typically involves "pure physical prehensions," that is, prehensions objectified by reproduced physical prehensions in the physical pole; and causal affection is a product of "hybrid physical prehensions" objectified solely by conceptual prehensions in the mental pole. The objectifications associated with hybrid prehensions entail concrete, real relations between (i) the actualization under way and (ii) the potentia belonging to the whole of the antecedent universe. The antecedence operative in hybrid prehensions is objective, given by the logical order governing the historical, genetic division of data in the mental pole; it is not relativistic, as given by the spatiotemporal order governing the spatiotemporal coordinate division of data in the physical pole.

The objective logical, historical ordering of events in the supplementary stage—reflected in the Ontological Principle, and also presupposed by the decoherence effect—precludes the possibility of purely symmetrical, external relations among actualities. Actualities, by the logical order, are always asymmetrically related such that there is only a one-way conditioning mediating them: The closed past conditions the open future, and never the reverse.

All this is entirely intuitive. There is no paradox or mystery involved in the birth of a child in California "immediately affecting" the potentia associated with the father in Hong Kong. His associated potentia are clearly—and from the standpoint of his description as an historical route of occasions, immediately—affected, though he may not know it yet. Whether one says the man "becomes a father" when his child is born—that is, "immediately"—or when he learns his child is born via some causal chain of communication bound by special relativity—is, on a superficial level, perhaps a just a matter of preference. But if the father is properly described most fundamentally as an historical route of actualizations of potentia—potentia that are regularly affected (to varying degrees) by antecedent actualizations everywhere in the universe—one sees the significance of quantum mechanical nonlocality.

The relativistic spatiotemporal limitation of the causal influence of one actuality A upon a subsequent actuality B requires that A lie in the past light cone of B. Any actualities lying outside the past light cone of B are thus causal contemporaries of A—that is, mutually causally independent, or symmetrically independent—according to the relativistic spacetime order of nature and its role in the primary stage/physical pole. But they are historically, and therefore asymmetrically, interrelated according to the logical order of nature and its role in the supplementary stage/mental pole. The nonlocality experiments of Aspect et al. discussed earlier exemplify this dichotomy, where the probability valuations pertaining to a measurement in B are demonstrably superluminally *affected* by an antecedent measurement in A; but they are not superluminally causally *influenced*. No communicative signal, for example, can be sent superluminally from A to B. In terms of the relativistic spatiotemporal order operative in the primary stage/physical pole, events in A and B are, in Whitehead's metaphysical scheme, mutual contemporaries and thus causally mutually independent. But in terms of the logical order of the supplementary stage/mental pole, events in A and B are asymmetrically interrelated according to this order, such that logically antecedent actualities always condition the potentia of subsequent quantum mechanical actualizations.

But by the dipolarity of concrescence, the logical, asymmetrical order operative in the mental pole also finds its *reflection* in the physical pole, in the form of the relativistically invariant spacetime intervals operative in this pole. It is true that these invariant intervals are always coordinately divided/decomposed into diverse, potential, relativistically noninvariant spacelike and timelike intervals, and that the phenomena of time dilation and length contraction associated with these potential decompositions are extremely intriguing. But these phenomena simply do not constitute the ultimate character of the relations among actualities, nor was it Einstein's belief that they did. The fact of the *objective invariance* of the parent spacetime intervals, which are logically *fundamental to* their potential relativistic decompositions, ought not be overshadowed by the relativistic relations among the latter. The theory of relativity is more importantly a theory of invariance. (Einstein had, in fact, originally named this theory "Invar009tentheorie." It was first referred to as "the theory of relativity" in a review written by Max Planck,[9] and no doubt the eventual popularity of that name had much to do with the conceptually fascinating phenomena of relativistic time dilation, length contraction, and so on. Einstein was pleased with neither the popular overemphasis of these phenomena nor the name "theory of relativity," and he supported the efforts of a group of colleagues who campaigned to change the name back to "the theory of invariance"—unfortunately to no avail.)

The exact manner in which Whitehead incorporates special relativity into his metaphysical scheme—a topic attended to in chapter 5—has been the subject of a great deal of debate, particularly as regards the correlation of his scheme with various interpretations of quantum mechanics. Hartshorne, for example, wrote: "By accepting [the theory of special relativity] as ultimate, Whitehead rendered the great doctrine of events as summing up the influences of the past distressingly ambiguous. For 'past' has no clear meaning in relativity physics."[10]

In describing his "Principle of Relativity," Whitehead wrote, "If we allow for degrees of relevance, and for negligible relevance, we must say that every actual entity is present in every other actual entity. The philosophy of organism is mainly devoted to the task of making clear the notion of 'being present in another entity.'"[11] "The actual world must always mean the community of all actual entities, including the primordial actual entity called 'God' and the temporal actual entities."[12] In the confines of a brief discussion, these quotations clearly give a good hint as to how Whitehead intended to incorporate Einstein's theory of relativity into his metaphysical scheme; but he also makes his position on the matter quite clear in his theory of spatiotemporal extension given in Part IV of *Process and Reality*, and discussed at length in chapter 5 of this book. As we have seen, the "degrees of relevance" pertain specifically to the processes of negative selection and integration in the supplementary stage as exemplified by the processes related to decoherence. It is therefore clear that in Whitehead's metaphysical scheme, the order of actualities in the universe, as this order pertains to prehensions in the process of concrescence, is as necessarily dipolar as the concrescence itself: The primary stage accords with the relativistic spacetime order, such that a datum lying inside the past light cone of a prehending subject is, in one sense, "more relevant" to the subject than a datum lying outside its past light cone. For this latter datum is a "contemporary" of the subject, and therefore causally noninfluential of it. This qualification, operative in the physical pole, derives jointly from (i) the relativistic spatiotemporal coordination of data by the subject—that is, its coordinate decomposition of the invariant spacetime intervals associated with *the* actual world into noninvariant timelike and spacelike intervals which contribute to the distinctiveness of *its* actual world; and (ii) the conditioning of this coordinate division by the constancy of the critical velocity c.

Nevertheless, such a datum may still be conceptually prehended by virtue of its antecedence in the logical order operative in the mental pole, and thus may affect the potentia integrated in this pole. In quantum mechanics, this affection is manifest experimentally by the nonlocal correlations discussed above, which—if they were attributed to some causal efficacy of the physical pole—would require a superluminal influence. And thus, the varying "degrees of relevance" to which Whitehead refers will be reflected in the valuated objectifications of these data as potentia in the supplementary phase of the prehending subject. According to the decoherence interpretations of quantum mechanics, these varying degrees of relevance determine which data will be dismissed as irrelevant environmental detail, and which data will be integrated into the mutually exclusive and exhaustive alternative outcome states.

Apart from the heuristic value of the decoherence interpretations of quantum mechanics in this regard—namely, as a mechanism demonstrative of the dipolar interrelations between the primary and supplementary stages, and indeed demonstrative of the physical function of this dipolarity—the subject of nonlocal interactions and their relationship to Whitehead's metaphysics has been infamously troublesome and confusing. Of Whitehead's incorporation of special relativity into his cosmological scheme, physicist Henry Stapp, for example, writes:

> Each of [Whitehead's] events prehends (and is dependent upon) not all prior events, but only the events of its own "actual world." The actual world of a given event is the set of all actual events whose locations lie in the backward light-cone of its own location. [In an] attempt to bring his ontology into conformity with the demands of relativity theory . . . Whitehead chose to reconcile his philosophic

aims with the empirical facts by imposing special ad hoc conditions on his basic ontology rather than allowing the empirical facts to follow from his philosophic principles. These ad hoc conditions are complicated, unnecessary, and apparently incompatible with the quantum facts.[13]

Stapp suggests a more coherent, modified Whiteheadian scheme, inspired by developments in the study of quantum nonlocality and intended to accommodate quantum mechanics. Stapp's proposed modification entails that "each event embodies all prior creation and establishes a new set of relationships among the previously existing parts. Thus each event embraces all of creation and endows it with a new unity."[14] But this is really no modification at all; it is simply Whitehead's own characterization of prehensions involved in integrations of potentia in the supplementary stage/mental pole. The reason for the confusion, however, is simple, for Stapp writes of his modified Whiteheadian theory:

> The theory proposed here is not exactly the one proposed by Whitehead. In the first place it ignores the mental aspects and concentrates instead on the space-time and momentum-energy aspects, in order to bring the theory into contact with theoretical physics.[15]

That a physicist would be inclined to disregard the processes of the mental pole in Whitehead's metaphysical scheme is understandable, which is why the equivalent term "supplementary stage," with its absence of any misleading implications of Cartesian dualism, is preferable in such syntheses. But to abandon the processes of the supplementary stage in an effort to apply Whitehead's cosmological scheme to quantum mechanics amounts to removing the engine from a car and attempting to use it for transportation regardless. The modern decoherence-based interpretations of quantum mechanics, with their conceptual and mechanical analogs to these supplementary-stage processes, evince this absolutely; for without these processes, the whole of the metaphysical scheme—and most certainly its application to quantum mechanics—is rendered totally incoherent. As suggested in chapter 5, there are *reasons* for the relativistic space-time order as it applies to causality—reasons pertaining to the privacy and publicity of actual occasions and nexūs of occasions as subjects and objects, reasons pertaining to the subjective aim of "balanced complexity," reasons pertaining to the derivation of freedom from the mutual independence of contemporaries, and so forth.

When the theory of relativity is interpreted as a theory of sheer subjectivity, as opposed to a theory of invariance, Einstein's important contributions do indeed seem wholly incompatible with Whiteheadian metaphysics. This might have been a factor in Hartshorne's enthusiastic endorsement of Stapp's modified Whiteheadian cosmological scheme, even though it wholly excises the supplementary stage and its operations in an attempt to conform Whitehead's philosophy to quantum mechanics. Of Stapp's modified Whiteheadian model, Hartshorne wrote:

> The process view as now revised is, in basic essentials, the simplest and most straightforward cosmology ever conceived that is compatible with what we now know about nature. . . . Philosophy can hardly abstain much longer from considering Stapp's startling idea, derived from Whitehead with help from Bell [and his theories pertaining to quantum nonlocality] that the four dimensions of space-time are our way of picturing relations obtaining, in terms of various types of influence, among instances of creative becoming whose well-ordered series of conditions and conditioned fall upon a single ultimate dimension. This dimension is far more like an ultimate time than an ultimate space, since it has a radical directionality, each member conditioning (when all types of influence are taken into account) all those coming "after" and none of those coming "before" it. It is time's one-way dependence, not space's symmetrical dependence or interdependence, that is the clue to reality in general or as such. Taken absolutely, space is, as so many philosophers have suspected, an illusion. This result is only what formal logic should have led us to expect.[16]

Here Hartshorne, understandably inspired by the implications of quantum nonlocality, relegates the relativistic spacetime ordering of physical prehensions in the primary stage to the status of epistemic illusion. But the fact that the modern decoherence-based interpretations of quantum mechanics perfectly exemplify the whole of Whitehead's metaphysical scheme, including the supplemental stage (indeed, *requiring* it) demonstrates that the relativistic space-time order of the primary stage need not be deleted from the scheme—nor should it ever have been, solely on the grounds of quantum nonlocality. For this nonlocality is causally noninfluential (neither signals, nor energy of any kind, can be transmitted superluminally). And yet nonlocal interrelations are real. They do exist, and they are most coherently and intuitively understood according to the

Whiteheadian metaphysical scheme. The decoherence-based inter-pretations demonstrate the usefulness of the dichotomy of the physi-cal ordering of the universe into two modes:

1. A relativistic spacetime order operative in the primary stage—the order to which causal efficacy must conform, and therefore the order which allows for mutually independent, symmetrically related, causally noninfluential contemporaries. This causal independence of spacelike-separated actualities is associated with the decoherence ef-fect, which eliminates (i.e., as negative prehensions) causally inde-pendent potentia[17] from further integration into the mutually exclusive, probability-valued alternative outcome states (i.e., "sub-jective forms") of the reduced density matrix. Since the decoherence effect takes place only by virtue of the logical ordering of actualiza-tions, however, the relativistic spacetime ordering of actualities oper-ative in the primary phase *presupposes* the asymmetrical logical ordering of the actualizations of potentia of the supplementary phase. This is a feature of the dipolarity of concrescence, as is the relativistic invariance of spacetime intervals in the physical pole as a reflection of the logical ordering of actualities in the mental pole.

2. A logical order characterizing the supplementary stage—the order to which the actualizations of potentia must conform. It is this order which permits the decoherence effect to occur at all, for it conditions the initial integration of prehended data as logically, asymmetrically ordered, thus establishing the asymmetrical relation-ship between past and future such that the former is always closed and the latter always open. Without this logical mediation of the punctuated actual and the fluid potential, the purely symmetrical temporal relations among spacelike-separated contemporaries im-plied by special relativity would prove wholly incompatible with Whiteheadian metaphysics, quantum mechanics, and the laws of thermodynamics.

It is significant that Hartshorne's abhorrence for the symmetrical external relations associated with some aspects of special relativity was generated largely out of theological concerns. The proposed dis-placement of relativistically noninvariant temporal and spatial order-ing operative in the physical pole by the logical, asymmetrical ordering transplanted from the mental pole, as suggested by Stapp, produces, according to Hartshorne, a great gain in coherence for the process view of reality, which was gravely compromised by taking relativity as the last word on the structure of space-time. . . . This change also simplifies, if it does not first make possible, the influence upon the world that Whitehead attributes to divine decisions. They only need be inserted between successive events in the ultimate series. Only those who know the troubles process philosophers have had in trying to insert divine influences into the world of mutually independent contemporaries know what a relief this doctrine affords.[18]

Because the theological implications of Whitehead's cosmological scheme lie outside the scope of this discussion, the function of God in this scheme and the exemplification of this function in quantum mechanics will be attended to only briefly here as a suggested topic for further investigation. In the context of the discussion thus far, the issue Hartshorne raises here seems to concern the imagined mechanism by which God interacts with the world, such that if this mechanism be causal, it is inferred that divine interaction would be somehow limited by the bounds of special relativity. This notion is, of course, problematic for a number of reasons. It would seem, though, that the concept of divine interaction with a world of logically ordered actualities and histories of actualities is entirely capable of accommodation by Whitehead's cosmology, via the processes of the supplementary stage, even without Stapp's proposed modification; indeed, in Whitehead's scheme, divine interaction is closely related to the function of pure potentia and is, in that regard, a requirement of the scheme.

The supposed problem of the physical interaction of God with an actual world ordered (in part) by relativistic spacelike and timelike intervals—like the closely related problem of nonlocality discussed earlier—disappears altogether when the processes of the supplementary stage are included in the metaphysical scheme as intended. The same thing can be said as regards the other four "problematic" conceptual quantum mechanical innovations enumerated by Shimony.

There are, however, two more important implications of Whitehead's metaphysical scheme, and its exemplification in quantum mechanics, which pertain to the concept of God. First is the relationship of this concept to the "subjective aim" toward effective complexity or "balanced complexity" as the *summum bonum* in Whitehead's cosmology. What role does God play in the mechanism that encourages (but does not determine) a balance between regularity and diversity in nature—or, at the quantum mechanical level in Whiteheadian terms, reproduction and reversion? Recent theoretical studies of complex adaptive systems in nature have led some to conclude that effective complexity is, in Whiteheadian terms, the "subjective aim" by which nature regulates herself without determining herself. The scientific and mathematical concepts involved in this regulation accord with Whitehead's Categoreal Obligations practically analogously. This is exemplified at the most fundamental physical level by quantum mechanics and its inherent concepts of regularity and diversity, reproduction and reversion, both qualitatively and quantitatively demonstrable.

Second is the role of God as primordial actuality in quantum mechanical cosmogonic models of *creatio ex nihilo,* such as the one proposed by Stephen Hawking and James Hartle.[19] Quantum mechanics describes the evolution of the state of a system of actualities always in terms of an initial state antecedent to the evolution, and a matrix of probable outcome states subsequent to and consequent of the evolution. Therefore, any model of quantum-mechanically described cosmogony—an inflationary universe model, for example— still requires a set of "initial conditions" or initial actualities at $t = 0$. For without these initial actualities, there is nothing that might evolve quantum mechanically, and no concrete physical spacetime structure in which such an evolution might operate. Though the abstract geometrical elements inherent in the morphological structure of spacetime are, for Whitehead, necessarily a priori and independent of actualities, a *purely* abstractive spacetime with no concrete physical character whatever simply cannot accommodate the metaphysical conception of concrescence, or the physical conception of quantum mechanical state evolution. For without data to prehend, there can be no concrescence. And without an initial actual state, there can be no quantum mechanical evolution to a final actual state. The abstract geometrical elements underlying the morphology of spacetime are, for Whitehead, potential forms of facts. Thus, even though these a priori forms do not *derive* from facts, apart from the facts, the forms are inoperative. Actuality cannot evolve from pure potentiality alone.

In their 1983 paper "The Wave Function of the Universe," Hartle and Hawking proposed a quantum mechanical model of cosmogony intended to function without the need of antecedent actuality—that is, without initial conditions of any kind, including spacetime structure itself. Quantum mechanics, by this model, describes the evolution of a state of a system of *sheer potentiality* into a state of actual facts constituting the nascent universe and its spacetime structure. This proposed initial state of sheer potentiality—often referred to by the term "quantum vacuum"—is describable mathematically as the empty set in set theory, and it is equated by Hartle and Hawking with the metaphysical concept of sheer nothingness. Thus, they claim, their model describes, via nothing more than quantum mechanics, the genesis of the universe as a quantum mechanical evolution from nothing. Hawking later clarifies the theological implications of this model by stating that it allows for the elimination of the notion of God as it applies to the creation of the universe.[20]

There are, however, a number of conceptual inconsistencies and even semantic equivocations that severely undermine Hartle and Hawking's claim that their cosmogonic model coherently describes *creatio ex nihilo ab nihilo*. And given that this model is a quantum mechanical one, it is not surprising that these inconsistencies are rooted in the same familiar ontological misapprehensions already described in this book, and clarified via the decoherence-based interpretations. Most basic is the attempt to escape the simple truth—which will forever restrain by the inescapable grip of logic—that quantum mechanics, by its necessary presupposition and anticipation of actualities, can never account for the existence of actualities. This pertains to the Hawking-Hartle cosmogonic model in two ways.

First, the initial state of the evolution, described as a state of sheer potentiality and thus a state of metaphysical "nothingness," nevertheless fulfills the *function* of antecedent actuality in this model. In this way, the model entails a conflation of actuality and potentiality similar to that which characterizes the matrix of outcome states in Everett's "many worlds" interpretation, wherein every potential outcome state is considered actualized. In the Hawking-Hartle model, it is not the consequent potential outcome states that are mischaracterized as actual, but rather the antecedent actual initial state that is mischaracterized as sheerly potential. This amounts to an attempt to equate the following two expressions of quantum mechanical state evolution—the first, a typical expression, followed by its supposed equivalent in the Hartle-Hawking model:

Typical: Quantum mechanics reveals a probability that actuality A at time $t = 0$ will become specified as actuality $A_{t+1} = 243.3$ at time $t + 1$ subsequent to and consequent of the interaction of A with some other actuality (an environmental actuality, a measuring apparatus actuality, etc).

Hartle-Hawking: Quantum mechanics reveals a probability that a potential actuality $?_? = ?$ (which is somehow at the same time referent to no actuality), will become specified as an actuality $A_{t+1} = 243.3$ at time $t + 1$ subsequent to and consequent of the interaction of potential actuality $?_?$ and another potential actuality $?_?$.

Clearly, these are not equivalent expressions. The concept of probability is so vitiated in the second expression that its coherent application to a quantum mechanical evolution is impossible. For what value is there in the statement: "I have predicted the probability that some possibility, referent to nothing, will become a ninety-two carat diamond"? Substitute the universe for the diamond, and Hawking's quantum mechanical cosmogony *ex nihilo ab nihilo* seems entirely inadequate to the gravity of its intent.

Furthermore, as evinced by the decoherence-based interpretations, a quantum mechanical evolution proceeds only by virtue of (i) necessary interrelations with actualities antecedent to the evolution; (ii) the necessary elimination, by negative selection, of a large number of potentia related to environmental facts. The necessity of the latter is an exemplification of Whitehead's Category of Transmutation in the supplementary stage of concrescence, and is the process by which superpositions of interfering potentia are eliminated, yielding a reduced density matrix of noninterfering mutually exclusive and exhaustive probability-valued outcome states. Simply, the lack of interrelations with actualities antecedent to a quantum mechanical evolution renders the evolution impossible. At the very most, an evolution of a system containing no actualities would yield a nonsensical, coherent superposition of alternative states, rather than a nascent universe with spatiotemporal order. But even that is too much to ask of such an evolution, for without antecedent actualities, there is nothing to evolve.

In quantum mechanics, a *potential* is strictly defined as a *possible specification of the actual;* without the actual, there is nothing to specify. "Actual" and "potential" are fundamental concepts in the ontological interpretation of quantum mechanics, such that neither is capable of abstraction from the other; clearly, then, the deletion of the concept "actuality" from the quantum mechanical definition of "potential" as "potential actuality" renders the latter term incoherent. It is certainly no longer the fundamental concept used in quantum mechanics. Nevertheless, of his proposed sheerly potential initial state of the quantum mechanical evolution of the universe, Hawking writes, "to ask what happened before the universe began is like asking for a point on Earth at 91 degrees north latitude; it just is not defined."[21] What is important, and overlooked here, is the "it" that is not defined—the "what" that evolved quantum mechanically to become the universe.

At first glance, the deletion of "actuality" from the quantum mechanical concept of "potential actuality" might seem acceptable, for semantically, the term "potential fact" as used in quantum mechanics is simply an example of an adjective modifying a noun. Abstracting the adjective from the pair does not render it conceptually nonsensical, nor does it even alter the meaning. Perhaps it is by an analogous conception that Hartle and Hawking conceived of their initial sheerly potential/nonactual initial state from which the actual universe purportedly evolves quantum mechanically. Clearly, though, the adjective "potential" in abstraction from the phrase "potential fact" only maintains its meaning because even in abstraction it tacitly presupposes the concept of "fact"—just as quantum mechanics does.

Second, the "nothingness" of Hawking and Hartle's proposed initial state of sheer potentiality—even if we were to grant that these *potentia ex nihilo* might evolve according to the Schrödinger equation though they refer to no antecedent actuality—evolves according to an ordered mathematical structure evinced by this equation. This complex order is utterly belied by popular terms such as "quantum foam" used to refer to this proposed initial state of sheer potentia, as though this state were some sort of chaotic froth. Sometimes "quantum vacuum" is used instead, and the fact that these two terms are interchangeable does not bode well for the models that use them. Joseph Życiński summarizes these first two objections thus:

The basic problem remains, however that in none of Hawking's models is the very notion of nothing (*nihilum*) accepted in the sense in which it was classically understood in a metaphysical description of *creatio ex nihilo*. When "nothing" denotes "something," the so-called "creation," in a physical sense of this term, can denote anything.[22]

It should be emphasized, however, that these quantum cosmogonies correctly suppose that quantum mechanics is best interpreted ontologically; indeed, James Hartle—cocreator of the model described above—is also a key contributor to the decoherence interpretations. The application of an ontologically interpreted quantum mechanics to cosmogony, as well as other areas of mutual significance to science, philosophy, and theology, not only should be encouraged; ultimately it cannot be avoided. And though quantum mechanics, by its fundamental presupposition of actualities, is not appropriate to the task of accounting for the origin of the universe, it is certainly well suited to the description of the evolution of the universe (once the facts of its existence are stipulated, as they are with any application of quantum mechanics).

Insofar as quantum mechanics exemplifies the coherent, logical, applicable, and adequate metaphysical scheme of Whitehead to an exceedingly fine level of analogous detail—both conceptually and in terms of their respective "mechanical" processes—it is clear that the many implications of this cosmology as they might apply to conversations among philosophy, science, and religion should be carefully explored. For it is only in the context of a coherent and logical metaphysical scheme that an ontological interpretation of quantum mechanics can be properly applied to matters such as cosmogony, which involve philosophy and religion as much as physical science. Apart from such a context, errors are likely to abound; and when these errors involve fundamental concepts, even the most complex and carefully thought out calculation is doomed to incoherence, which, according to Whitehead, "is the arbitrary disconnection of first principles."

> In modern philosophy Descartes' two kinds of substance, corporeal and mental, illustrate incoherence. . . . The requirement of coherence is the great preservative of rationalistic sanity. But the validity of its criticism is not always admitted. If we consider philosophical controversies, we shall find that the disputants tend to require coherence from their adversaries, and to grant dispensations to themselves.[23]

In modern physics, one might conclude the same thing about Bohr's Principle of Complementarity if it were presented as an ontology rather than just an epistemology. (Though, as mentioned before, any epistemological sanction entails an ontological generalization.)

The application of the decoherence-based interpretations of quantum mechanics to the study of complexity in nature, where the former is seen as a fundamental exemplification of the latter, is, like quantum cosmogony, an area of inquiry that entails both the philosophy of science and the philosophy of religion. The contextualization of quantum mechanics in terms of the Whiteheadian metaphysical scheme is commended in this conversation as well—especially since the tendency toward complexity in nature is a fundamental feature of Whitehead's philosophy, as discussed earlier. Whitehead's repudiation of mechanistic materialism is also a repudiation of its correlate characterization of the universe as a cold realm of mechanical accidents from which our purportedly illusory and sheerly subjective perceptions of purpose and meaning are, according to certain views, thought to derive. To repeat the words of Jacques Monod, the Nobel-laureate biochemist, already quoted in the introduction of this book: "Man knows at last that he is alone in the universe's unfeeling immensity, out of which he emerged only by chance."[24] In sharp contrast, by Whitehead's cosmology, the universe is instead characterized as a fundamentally complex domain, a nurturing home for a seemingly infinitely large family of complex adaptive systems such as ourselves.

The paradigm of speculative philosophy is particularly suited to such conversations among philosophy, science, and religion, so long as the logical limitations of each of the three perspectives are explicitly understood. The role of science is to transmute accurate observations into hypothetical generalizations, with the understanding that this constitutes the beginning of the process and not the end; quantum mechanics, for example, logically cannot be used to *account for* cosmogony or, by the same logical requirement, to account for creative indetermination, free will, and so forth—though it can be used to *describe* aspects of all of these. Similarly, speculative philosophy should have little significance to matters of faith in science and religion conversations, since the great value of faith lies in the notion that its object need not be demonstrable—certainly not scientifically. The religious significance of such conversations, then, is best understood as a potential supplement to faith and not a substitute for it. There ought not, in other words, be anything threatening to either scientific, philosophical, or religious principles in these conversations. At the same time, it is not the task of science, philosophy, or religion to explain away those concepts or actualities that each necessarily presupposes; it is not the task because it is not logically possible. Attempts to apply science to such an endeavor, as evinced by the spirit of Hawking's quantum cosmogonic model, for example, are best avoided if the often delicate but ever more important dialogue among philosophy, science, and religion is to prove fruitful.

In the words of C. S. Lewis, which ought be heeded by all conversation partners:

> You cannot go on "explaining away" for ever: You will find that you have explained explanation itself away. You cannot go on "seeing through" things for ever. The whole point of seeing through something is to see something through it. It is good that the window should be transparent, because the street or garden beyond it is opaque. How if you saw through the garden too? It is no use trying to "see through" first principles. If you see through everything, then everything is transparent. But a wholly transparent world is an invisible world. To "see through" all things is the same as not to see.[25]

NOTES

1. Alfred North Whitehead, *Process and Reality: An Essay in Cosmology, Corrected Edition*, ed. D. Griffin and D. Sherburne (New York: Free Press, 1978), 6.

2. Ibid., 285.

3. Heisenberg's formulation of quantum mechanics in terms of matrix mechanics (as opposed to the later and equivalent "wave mechanics" formulation of Schrödinger) was published in 1925 in a paper entitled "Quantum-Theoretical Re-interpretation of Kinematic and Mechanical Relations."

4. Whitehead, *Process and Reality*, 117.

5. A. Shimony, "Search for a Worldview Which Can Accommodate Our Knowledge of Microphysics," in *Philosophical Consequences of Quantum Theory: Reflections on Bell's Theorem*, ed. J. Cushing and E. McMullin (Notre Dame, Ind.: Notre Dame University Press, 1989), 27.

6. Max Born, "Bemerkungen zur statistischen Deutung der Quanten-

mechanik," in *Werner Heisenberg und die Physik unserer Zeit*, ed. F. Bopp (Braunschweig: F. Vieweg, 1961), 103–118.

7. Whitehead, *Process and Reality*, 5–6.

8. Ibid., 4–5.

9. Irving M. Klotz, "Postmodernist Rhetoric Does Not Change Fundamental Scientific Facts," *The Scientist*, 22 July 1996.

10. Charles Hartshorne, "Bell's Theorem and Stapp's Revised View of Space-Time," *Process Studies* 7, no. 3 (1977): 184.

11. Whitehead, *Process and Reality*, 50.

12. Ibid., 65.

13. H. Stapp, "Quantum Mechanics, Local Causality, and Process Philosophy," *Process Studies* 7, no. 3 (1977): 176–177.

14. Ibid., 175.

15. Ibid., 175.

16. Hartshorne, "Bell's Theorem," 187.

17. Because these are potentia that are causally independent relative to the system measured and the measuring apparatus, they are necessarily "environmental" to the system and apparatus, and therefore eliminated in the trace-over of the pure state density matrix.

18. Hartshorne, "Bell's Theorem," 187.

19. James Hartle and Stephen Hawking, "The Wave Function of the Universe," *Physical Review D* 28 (1983): 2960–2975.

20. R. Weber, *Dialogues with Scientists and Sages: The Search for Unity* (London: Routledge, 1986), 212.

21. Stephen Hawking, "Quantum Cosmology," in *Three Hundred Years of Gravitation*, ed. S. Hawking and W. Israel (Cambridge: Cambridge University Press, 1987), 651.

22. Joseph Życiński, "Metaphysics and Epistemology in Stephen Hawking's Theory of the Creation of the Universe," *Zygon* 31, no. 2 (1996): 275.

23. Whitehead, *Process and Reality*, 6.

24. Jacques Monod, *Chance and Necessity: An Essay on the Natural Philosophy of Modern Biology*, trans. Austryn Wainhouse (New York: Vintage Books, 1972).

25. C. S. Lewis, *The Abolition of Man, or Reflections on Education with Special Reference to the Teaching of English in the Upper Forms of Schools* (London: Macmillan, 1947), 91.

WORKS CITED

Aspect, A., J. Dalibard, and G. Roger. "Experimental Test of Bell's Inequalities Using Time-Varying Analyzers." *Phys. Rev. Lett.* 44 (1982): 1804–1807.

Bell, John. "Against Measurement." In *Sixty-Two Years of Uncertainty: Historical, Philosophical, and Physical Inquiries into the Foundations of Quantum Mechanics*, ed. Arthur I. Miller, 17–31. New York: Plenum, 1990.

————. "On the Einstein Podolsky Rosen Paradox." *Physics* 1, no. 3 (1964): 195–200.

————. *Speakable and Unspeakable in Quantum Mechanics*. Cambridge: Cambridge University Press, 1987.

Bohm, David. "A Suggested Interpretation of the Quantum Theory in Terms of 'Hidden' Variables." *Phys Rev.* 85 (1952): 166–193.

————. "Time, the Implicate Order, and Pre-space." In *Physics and the Ultimate Significance of Time*, ed. David R. Griffin, 177–208. New York: State University of New York Press, 1986.

Bohm, David, and B. J. Hiley. *The Undivided Universe: An Ontological Interpretation of Quantum Theory*. London: Routledge, 1993.

Bohr, Niels. *Atomic Physics and Human Knowledge*. New York: Wiley, 1958.

————. *Atomic Theory and the Description of Nature*. Cambridge: Cambridge University Press, 1934.

————. "Discussion with Einstein." In *Albert Einstein: Philosopher-Scientist*, ed. Paul Arthur Schilpp, 199–242. Evanston: Library of Living Philosophers, 1949.

Born, Max *Atomic Physics*, 8th ed. New York: Dover, 1989.

————. "Bemerkungen zur statistischen Deutung der Quantenmechanik." In *Werner Heisenberg und die Physik unserer Zeit*, ed. F. Bopp, 103–118. Braunschweig: F. Vieweg, 1961.

————. *Natural Philosophy of Cause and Chance*. Oxford: Oxford University Press, 1949.

Eastman, Timothy E. "Micro- to macroscale perspectives on space plasmas." *Phys. Fluids* B5 (7): 2671–2675 (1993).

Einstein, Albert. "Zur Elektrodynamik bewegter Korper." *Annalen der Physik* 17 (1905): 892. Trans. Arthur I. Miller, "On the Electrodynamics of Moving Bodies," appendix in *Albert Einstein's Special Theory of Relativity*, 393. Reading, Mass.: Addison-Wesley, 1981.

Everett, Hugh. "'Relative State' Formulation of Quantum Mechanics." *Rev. Mod. Phys.* 29 (1957): 454–462.

Folse, Henry. "The Copenhagen Interpretation of Quantum Theory and Whitehead's Philosophy of Organism." *Tulane Studies in Philosophy* 23 (1974): 32–47.

Freedman, S., and J. Clauser. "Experimental Test of Local Hidden Variables Theories." *Phys. Rev. Lett.* 28 (1972): 938–941.

Frieden, B. Roy. *Physics from Fisher Information*. Cambridge: Cambridge University Press, 1999.

Geach, P. T. *God and the Soul*. London: Routledge, 1969.

Gell-Mann, Murray. *The Quark and the Jaguar: Adventures in the Simple and the Complex*. New York: W. H. Freeman, 1994.

Gell-Mann, Murray, and James Hartle. "Strong Decoherence." *Proceedings of the Fourth Drexel Symposium on Quantum Non-integrability: The Quantum-Classical Correspondence*. Drexel University, September 1994.

Ghirardi, G. "Letters." *Physics Today* 46, no. 4 (1993): 15.

———. "Macroscopic Reality and the Dynamical Resolution Program." In *Structures and Norms in Science*, ed. M. L. Dalla Chiara et al. Dordrecht, Netherlands: Kluwer Academic, 1997.

Ghirardi, G., A. Rimini, and P. Pearle. "Old and New Ideas in the Theory of Quantum Measurement." In *Sixty-Two Years of Uncertainty*, ed. A. Miller, 167–193. New York: Plenum, 1990.

Griffiths, Robert J. "Consistent Histories and the Interpretation of Quantum Mechanics." *Stat. Phys.* 36 (1984): 219–272.

———. *Consistent Quantum Theory*. Cambridge: Cambridge University Press, 2002.

Hartle, James, and Stephen Hawking. "The Wave Function of the Universe." *Physical Review D* 28 (1983): 2960–2975.

Hartshorne, Charles. "Bell's Theorem and Stapp's Revised View of Space-Time." *Process Studies* 7, no. 3 (1977): 183–191.

Hawking, Stephen. "Quantum Cosmology." In *Three Hundred Years of Gravitation*, ed. S. Hawking and W. Israel, 631–651. Cambridge: Cambridge University Press, 1987.

Heelan, Patrick, S.J. *Quantum Mechanics and Objectivity*. The Hague: Martinus Nijhoff, 1965.

Heisenberg, Werner. "The Development of the Interpretation of the Quantum Theory." In *Niels Bohr and the Development of Physics*, ed. Wolfgang Pauli, 12–29. New York: McGraw-Hill, 1955.

——. *Physics and Philosophy*. New York: Harper Torchbooks, 1958.

Heitler, Walter. "The Departure from Classical Thought in Modern Physics." In *Albert Einstein: Philosopher-Scientist*, ed. Paul Arthur Schilpp, 179–198. Evanston: Library of Living Philosophers, 1949.

James, William. *Pragmatism and the Meaning of Truth*. Cambridge: Harvard University Press, 1978.

Klotz, Irving M. "Postmodernist Rhetoric Does Not Change Fundamental Scientific Facts." *The Scientist*, 22 July 1996.

Lewis, C. S. *The Abolition of Man, or Reflections on Education with Special Reference to the Teaching of English in the Upper Forms of Schools*. London: Macmillan, 1947.

Lorentz, H. A., A. Einstein, H. Minkowski, and H. Weyl. *The Principle of Relativity: A Collection of Original Memoirs on the Special and General Theory of Relativity*. New York: Dover, 1952.

Lucas, George. *The Rehabilitation of Whitehead: An Analytic and Historical Assessment of Process Philosophy*. New York: State University of New York Press, 1989.

Monod, Jacques. *Chance and Necessity: An Essay on the Natural Philosophy of Modern Biology*. Trans. Austryn Wainhouse. New York: Vintage Books, 1972.

Omnès, Roland. *The Interpretation of Quantum Mechanics*. Princeton, N.J.: Princeton University Press, 1994.

Popper, Karl. *Quantum Theory and the Schism in Physics*. New Jersey: Rowman and Littlefield, 1956.

Schrödinger, Erwin. "Die gegenwartige Situation in der Quantenmechanik." *Naturwissenschaften* 23 (1935): 807–812, 823–828, 844–849. English translation: John D. Trimmer, *Proceedings of the American Philosophical Society* 124 (1980): 323–338.

Shimony, Abner. "Quantum Physics and the Philosophy of Whitehead." In *Boston Studies in the Philosophy of Science*, ed. R. Cohen and M. Wartofsky, 2:307–330. New York: Humanities Press, 1965.

——. "Search for a Worldview Which Can Accommodate Our

Knowledge of Microphysics." In *Philosophical Consequences of Quantum Theory: Reflections on Bell's Theorem*, ed. J. Cushing and E. McMullin, 25–37. Notre Dame, Ind.: Notre Dame University Press, 1989.

Stapp, Henry P. *Mind, Matter, and Quantum Mechanics*. Berlin: Springer-Verlag, 1993.

———. "Quantum Mechanics, Local Causality, and Process Philosophy." *Process Studies* 7, no. 3 (1977): 173–182.

Von Neumann, John. *Mathematical Foundations of Quantum Mechanics*. Princeton, N.J.: Princeton University Press, 1955.

Weber, R. *Dialogues with Scientists and Sages: The Search for Unity*. London: Routledge, 1986.

Whitehead, Alfred North. *The Concept of Nature*. Cambridge: Cambridge University Press, 1920.

———. *Modes of Thought*. New York: Macmillan, 1938.

———. *Process and Reality: An Essay in Cosmology, Corrected Edition*. Ed. D. Griffin and D. Sherburne. New York: Free Press, 1978.

———. *Science and the Modern World*. New York: Free Press, 1967.

Whitehead, Alfred North, H. Carr, Wildon Carr, and R. A. Sampson. "Relativity, Logic, and Mysticism." *Aristotelian Society, Supplementary* 3 (1923): 34–41.

Wigner, E. P. "Remarks on the Mind-Body Question." In *The Scientist Speculates: An Anthology of Partly-Baked Ideas*, ed. Irving John Good, 284–301. New York: Basic Books, 1962.

Żurek, Wojciech. "Decoherence and the Transition from the Quantum to the Classical." *Physics Today* 44, no. 10 (1991): 36–44.

———. "Letters." *Physics Today* 46, no. 4 (1993): 84.

Życiński, Joseph. "Metaphysics and Epistemology in Stephen Hawking's Theory of the Creation of the Universe." *Zygon* 31, no. 2 (1996): 269–284.

INDEX